A Natural History of
BRITAIN and
IRELAND

ERIC SIMMS

ILLUSTRATED BY ROBERT GILLMOR

J. M. DENT & SONS LTD
LONDON MELBOURNE TORONTO

'I will walk the meadows, by some gliding stream, and there contemplate the lilies that take no care, and those very many other various little living creatures that are not only created, but fed, man knows not how, by the goodness of the God of Nature.'

IZAAK WALTON

First published 1979
© Eric Simms 1979

British Library Cataloguing in Publication Data

Simms, Eric
A natural history of Britain and Ireland.
1 Natural history – Great Britain
I Title
574.941   QH137

ISBN 0-460-04372-2

Set in 11/12pt VIP Baskerville by
DP Media Ltd, Hitchin
Printed in Great Britain by
Billing & Sons Ltd
Guildford, London & Worcester
for J. M. Dent & Sons Ltd
Aldine House   Welbeck Street   London

Contents

A Simple Guide to the Rocks

Era	Period	Approximate Age in Years	Features	Distribution
Quaternary	Historic		Present flora	Shingle beaches, estuary mud.
	Pre-Historic		Peat	Fenland alluvium. Peat. Glaciation with drift and boulder clay over much of country.
	Pleistocene	1,000,000	Glacial and interglacial deposits	Red Crag, Cromer Beds, Chalky boulder clay.
Tertiary	Pliocene		Temperate flora Britain's structure complete	Shelly sands and crag. Kent and E. Anglia.
	Miocene	25,000,000		Little in Britain.
	Oligocene			Limestone clays of Hampshire Basin.
	Eocene	70,000,000	Tropical swamps	London clay, Bagshot sands, Hampshire Basin and I.O.W. Volcanic rocks of Antrim, Mull, Skye.
Secondary or Mesozoic	Cretaceous		Flowering plants, early mammals and dinosaurs	Chalk (Dorset to Yorks), Gault and Greensand (around Weald, in I.O.W. and along chalk above). Sands and clay of Kent/Sussex Weald.
		135,000,000		
	Jurassic		Dinosaurs	Clays, limestones from Lyme Regis to N. Yorks.
		180,000,000		
	Triassic	225,000,000	Reptiles and brachiopods	New Red Sandstone and Keuper marl. N.E. Ireland. From Lower Severn to Lancs, Solway Firth, Yorks, Durham, E. Devon, Notts.
Primary or Palaeozoic	Permian	270,000,000	Amphibians	Conglomerates and red sandstone of Eden Valley and Devon. Magnesian limestone.
	Carboniferous	350,000,000	Ferns, Lycopods	Coal Measures, Millstone Grit, Mountain Limestone.
	Devonian	400,000,000	Fossil fish	Old Red Sandstone.
	Silurian		Earliest land plants Fish, corals, crinoids	Southern Uplands of Scotland, Parts of N.E. Ireland, Southern Lake District, N. and central Wales.
		440,000,000		
	Ordovician	500,000,000	Graptolites	Volcanic and sedimentary. Parts of Southern Uplands, Ireland, much of Lake District, parts of Wales and Shropshire.
	Cambrian	600,000,000	Trilobites, molluscs	Quartzites and Durness limestone of N. Scotland. Parts of N. Wales and Ireland.
Pre-Cambrian or Archaean	Pre-Cambrian	4,500,000,000?	Pre-fossils	Sedimentary Torridonian Sandstone of N.W. Scotland. Large areas of Archaean in Highlands and Islands of Scotland. Lewisian gneiss. Moine schists, in N. Scotland. Metamorphic rocks in S.W. Highlands, and Ireland. Volcanic rocks in Borders of Wales and Anglesey.

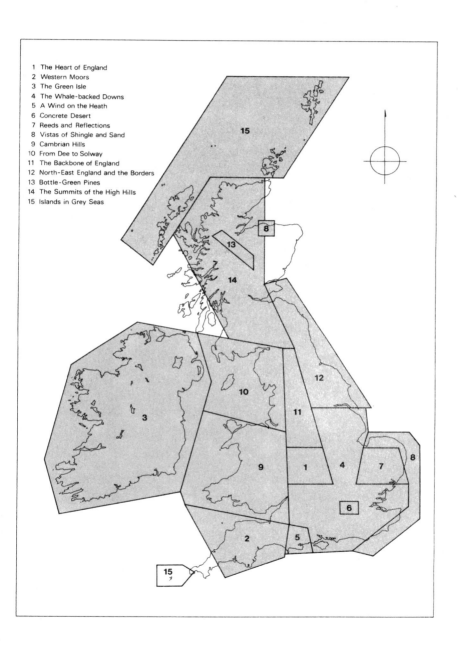

1 The Heart of England
2 Western Moors
3 The Green Isle
4 The Whale-backed Downs
5 A Wind on the Heath
6 Concrete Desert
7 Reeds and Reflections
8 Vistas of Shingle and Sand
9 Cambrian Hills
10 From Dee to Solway
11 The Backbone of England
12 North-East England and the Borders
13 Bottle-Green Pines
14 The Summits of the High Hills
15 Islands in Grey Seas

To the memory of GILBERT WHITE

Preface

In August 1971 I drove from London north to the town of Peterhead some 550 miles away. With me was my eleven-year-old son David and from the moment when we joined the M1 motorway, barely half a mile from our home at Dollis Hill in north-west London, he kept a written record of what we saw on our long journey – rocks and trees, farms, houses and wildlife. The first few miles of the motorway to Watford lay on the same deposits of Eocene clay which formed my own garden. We found ourselves running between lines of dark oily-green elms and oak trees, storm-slashed cornfields, green pastures and thorn hedges with blackbirds and magpies. Near Dunstable the landscape began to change, for now we were running on the chalk which dusted the fields with white, formed grey-white cuttings, sported calcicolous flowers and bore fewer trees, and the countryside became hillier and more rolling. Near Woburn we met the Cretaceous greensand with its light dry soil and scattered conifers, and here we saw the first sand martins, whose nesting tunnels are easily driven into the soft deposits. The belt of greensand was narrow and soon we were committed to the oolite of Bedfordshire and Northamptonshire which sweeps up from Portland Bill to Whitby and the parallel crescent of the lias in Leicestershire. This was an undulating region of farming land and grasslands on which large numbers of starlings, rooks and jackdaws were searching for worms and leatherjackets, and there were assemblies of lapwings and occasional small coveys of partridges. A kestrel was sitting on a road sign near the M6 linkway and soon others could be seen hovering above the motorway embankments.

On northwards we travelled over the Keuper marl and sandstone west of Leicester up to Nottingham, over flat farming country with scattered woods; here the Coal Measures, magnesian limestone and Bunter sandstone all met. The route was often astride the long tongue of magnesian limestone that reaches up to Richmond and we saw characteristic grey cuttings in the roadside, and sometimes pinker banks of the Bunter deposits. Much of the scenery was flat farming land broken by occasional long slopes and river valleys running across our path at right angles. Collieries showed up in the distance and there were duck circling above Fairburn Ings – flooded land not far from Ferrybridge. We spared a glance for the Devil's

Arrows on the right near Boroughbridge – three monoliths quarried at least seven miles from their present site – and viewed the Pennines to the west and the Hambleton Hills to the east. At Scotch Corner we turned off the A1 for Pearcebridge and the A68 for West Auckland, Toft Hill and Tow Law set on the Coal Measures, Carboniferous limestone and Millstone Grit. The land was more rugged with houses of grey-brown stone or brick, open-cast workings and mines, slag heaps and stone walls. There were ash trees which grew well on the limestone, birch scrub and marginal land. There was moorland on the unyielding grit with late pipits, curlews and larks – this was a mixed northern land often with wide views from the high ridges which carried the road. Rushy moors gave way to valleys with ash trees and sycamores, sessile oaks and smoking chimneys – and here starlings, linnets and pied wagtails were at home in this amalgam of countryside and industry. We reached Corbridge where we crossed the Tyne and Hadrian's Wall for a long drive over the Carboniferous limestone. Common gulls were flying above the moors and curlews – appropriate symbols of the Northumberland National Park – were gathering on the hills. Jackdaws frequented the villages and large conifer forests fringed the way. Pied wagtails were everywhere. Then came some Silurian rocks and the Old Red Sandstone of the Borders with heath, moor and valley oakwoods, while in the distance we could see the distant Cheviot. Beyond Galashiels meadowsweet and huge umbellifers flowered along the road and we saw many rooks, chaffinches and a party of migrant ring ouzels. Woodpigeons were singing and we could hear the autumn song of robins.

After a night in Peebles set amongst bare Silurian hills David and I continued our journey across heathy moors and a somewhat devastated landscape to Stirling, crossing Ordovician deposits, Old Red Sandstone, Coal Measures, Grit and Carboniferous limestone in fairly quick succession. On the lower Old Red Sandstone the countryside became more wooded, and very pleasing. After crossing the Highland Boundary Fault near Dunkeld and the coniferous woods of Blairgowrie we entered a darker mountainous land of Dalradian schists and granites – this was a region of curlew, wheatears, red grouse and meadow pipits. Buzzards and ravens greeted us in this high northern scenery and there was a chill in the air. From Royal Deeside with its sessile oakwoods and Scots pine forests echoing with woodpigeon song we swung towards the coast through Strathdon, rich in mountain hares, across more Dalradian rocks and under the long granitic bluff of Bennachie 'where the Gadie rins'. We followed the granite cliffs and sandy dunes of the Buchan coast with their seabirds and rose-pink fishing towns where the herring gulls wailed and screamed. Our transect of Britain was over but we had crossed many different geological belts and were left with a kaleidoscopic impression of rich farmlands, grazings, high grassy or heathery moors, mountain ranges, oakwoods and conifer plantations, sandy heaths, river gorges, marshy pools, sea cliffs and sand dunes, collieries, steelworks, towns and villages, isolated crofts and farmsteads. The journey had revealed the great variety and fascination of the British countryside in even a small part of the world, and the wildlife too is just as varied and fascinating. During the last war I spent 225 hours flying over the British Isles and was left with a similar impression of variety in a small region – something that many foreigners envy us who live on those islands.

Spring, for example, means different things to different people. For those who live by some southern forest, wood or copse it may be the wild song of the mistle thrush or the chiming notes of the chiffchaff back from Africa. For the farm worker it may be the mad rush of hares, pussy willow, the skylark declaiming his passionate message over the fields or a lapwing tumbling and falling about the sky in his spring ecstasy. For suburban man on his way to the station in the morning it may be the swelling chorus of blackbirds and house sparrows. The hill shepherd walking with his dog across the moors of northern Britain or Wales may think spring heralded for him by the bubble of the curlews or the strange sounds of the cock red grouse courting his hen. For the marshman digging for bait on his open estuary it may be the soft spring display of the ringed plover, the yodelling of a redshank or the scream of the first migrating tern. The crofter on Fair Isle may feel that it is the return of the piratical skuas while the fishermen in their seine net boats may pause as they return to harbour to listen to the growing cacophony from the seabird city on the cliff above their heads.

The countryside of the British Isles of which I speak is a dynamic one, changing with the times and reflecting the long and chequered history of man. Within historical times the wolf and the bear have been exterminated; the last wolf was killed in Scotland in 1743 and in Ireland perhaps as late as the 1760s, while the bear – the greatest British carnivore of the historic era – disappeared as long ago as the ninth or tenth centuries. The elk went about 1300, the beaver around 1500 and the wild boar, which finally lost its true identity in a mixed population of wild boars and escaped domestic pigs, may have survived as late as the seventeenth century. The reindeer also went in the middle of the twelfth century but it has been reintroduced into the Cairngorm Mountains. We have also lost the great auk, the bustard, the crane, the spoonbill, the Kentish plover and the large

13

copper butterfly, but we have regained the bittern, osprey, Savi's warbler, black-tailed godwit and ruff and gained the collared dove, redwing, fieldfare and Cetti's warbler. Often without forethought man has also let free or allowed to escape many creatures in Britain. Just think of wild goats, muntjacs, rabbits, grey squirrels, black and brown rats, coypus, mink, little owls, pheasants, red-legged part-ridges, ring-necked parakeets, carp, marsh frogs and others. The story of Britain and Ireland from the point of view of their natural history is one of continuous change and this is part of its fascination.

Two thousand years ago some 60 per cent of Britain was covered with forest, and trees in the mountains even reached a height of two thousand feet or more. Today 80 per cent is agricultural land with a largely eighteenth-century character. The original tree cover which had been determined by the influence of the underlying rocks, soil and climate was largely removed by man for charcoal, house and shipbuilding, fencing and fuel. Much of the deforested land has been replanted with carpets of alien conifers but today the need for hardwoods and mixed planting has been more widely recognized. The farmlands are a significant haven for wildlife, for many wood-land species found shelter on them and in the hedgerows, but the latter have often been removed to create new prairie-like fields. We have built towns and cities and made suburbs which for some woodland birds have proved more favourable than the woods from which they originally came. We have drained fenlands and marshes and have tended to make everything uniform and tidy, whereas it is essential to keep the widest variety of habitats and preserve their inhabitants. Land loss in lowland Britain to different forms of development seems inevitable and it has been running at the rate of some sixty thousand acres a year. How much of the land in the British Isles is truly natural – that is, untouched by man? The answer is very little indeed! Perhaps only on the summits of the highest mountains, on a few sea cliffs or islets or in a few fragments of inaccessible hillside woodland can the term 'natural' have any application at all. The rest of our countryside has been shaped, ploughed, managed, cultivated, cut, burned, planted or in some way influenced by man.

And what about the countryside itself? At the end of the eighteenth century with a population in England and Wales of around 8½ million people nearly 80 per cent lived in what we call the country. Now at least 80 per cent of that population live in towns and cities. As the nineteenth century progressed the lives of most people became for the first time divorced from the country and out of contact with nature. Since then some people have looked with indifference or even actual hatred at our countryside. From about 1870 this attitude has been counterbalanced by the outward movement of townspeople in a search for cleaner air and spiritual refreshment. With the arrival of the twentieth century came a revolution in cheap transport and the urban world exploded into the countryside. Into the green fields and woods came people with awakened sensitivities and a desire to learn, as well as red-roofed suburbia, stark concrete roads and motorways, airfields, massive structures of the new technological age, reeking rubbish dumps, hideous hoardings, new high ironmongery across the land and toxic chemicals, vandalism and lack of understanding. A journey across these islands reveals the acres of desert land, decrepit camps, derelict industrial sites, sterile slag heaps and abandoned rubbish, cars and other rejectamenta of our society.

It has been said that any country has in the end only two ultimate assets – its people and its land. The great problem of now and tomorrow is how to reconcile the many different needs and demands upon both when our land resources are so small and our population so high. Prosperity is dependent upon industrial growth, and sentiment also must not stand in the way of man's progress, but the problem of how we use or misuse our countryside is a pressing one. Should copper, China clay and fluorspar be mined in National Parks? Do planning authorities and conservationists establish mutual trust between themselves? There is now a much greater awareness of the problems and there are Government bodies and agencies, conservation and protection societies, Trusts and other organizations to monitor the changes and to try and reconcile productive land use with our enjoyment of it. There are some places where 'access in excess' could degrade the habitat which people come to see, and this has happened at Sennen Cove, on Box Hill and Cairn Gorm. There is also a conflict between the pure conservationist who detests the sight of another person in a particular environment and the other sort who want the beauty and variety of nature to be shared among as many people as possible. I have seen during my own membership the total membership of the Royal Society for the Protection of Birds rise from five thousand to over a quarter of a million. It is clear that not all the members could visit the same reserve in one day for obvious reasons and so control has to be exercised for the benefit of the wildlife. Perhaps we still have to learn to care enough about the future of our countryside, since we hold it in trust for coming generations. Personally I cannot improve upon the late Professor C. E. M. Joad's observations in 1946: 'The experiences which intercourse with nature and solitude bring are, indeed, valu-

15

able; a life which embodies them will be fuller and richer than a life which does not; and, if the community of the future denies them to its citizens, it will have been content to demand for them less than the best.' There are few places in the world where man and nature have co-operated so well as in the British Isles.

Fifty years ago I lay one August day on a sunny downland slope in Sussex, breathing in the aromatic sweet scent of thyme and marjoram, listening to a late skylark singing far above my head and looking out across a green and pleasant world – glowing in the sun. Grasshoppers chirped close to my resting place and humblebees droned away in the background, bumbling from hogweed clump to the next domed umbellifer. For the first conscious time in my life – and I was not quite ten years old – I experienced an overwhelming rush of feeling, almost a physical spasm, of love for the countryside of Britain with its astonishing beauty, variety and appeal. I have never lost that sense of wonder and I have been fortunate enough to spend much of my life investigating and marvelling at it, tape-recording its sounds, writing about it, directing films and producing television programmes about it as well as its human occupants and its wildlife. Since January 1952 when I started *The Countryside* programme which is still being broadcast regularly on Radio 4 I have travelled thousands of miles to gather material for its continuing editions. Other radio and television work has taken me across the British Isles from the Isles of Scilly to Shetland, from Kent to the Butt of Lewis, and from the Farnes to the Aran Islands in Galway Bay. Research for other books and the perpetual need for change and adventure have led me to many interesting parts of our islands.

In this book I have endeavoured to write about the varied nature of the countryside as seen through the eyes, and experienced with the hearing, of one naturalist who has travelled widely, and I hope neither obtusely nor unsympathetically. I can only hope that my readers will be able to share perhaps something of my feelings of wonder and some of my own experiences of the remarkably varied and fascinating countryside which we are privileged to know in the British Isles.

Each chapter is devoted to a different geographical area or ecological habitat in the British Isles and the map on page . . . shows the distribution of the various chapters. Some reveal the different facets of a region such as the midlands of England, Ireland or Wales, or a special habitat such as the heathlands of Hampshire and Dorset, the fenlands of eastern England, the old Caledonian pine forest and the northern isles. In this way it is hoped to show almost every facet of the countryside as the traveller may see them.

1. The Heart of England

For the naturalist early morning is surely the best time in which to see Nature with her defences down and to catch her off-guard. Many of my most rewarding experiences in the countryside have occurred just after dawn when the air is often still and man's intrusive noises – the sounds of traffic, trains and aircraft – are perhaps at their least compelling. This is certainly true of what I have chosen to call the heart of England – that undulating plain of grassland, small fields hedged about with thorn, oak, ash and dying elm, patches of woodland, farmstead and village that is centred on Banbury, with its meat market and industry, and reaches north to Rugby, south to Oxford, east to Aylesbury and west to the Cotswolds. But what kind of countryside is it?

Once it was overlain by muddy or iron-rich shallow seas which in Jurassic times determined the clays and sands of the lias and inferior oolite that, like strong horizontal muscles, uphold the heart of England. The liassic rocks fall into three main divisions – the lower lias, the middle and the upper. The upper lias gives rise to light brown clayey soils used for grassland and general arable farming; the middle arose from shallow-water limestones, which form the local hills, while the inferior oolite, composed of miniature spheres of calcium carbonate around a nucleus and shaped like the eggs of fish, occurs above the lias, capping the hills and helping to preserve them by its porosity and resistance to erosion. Another kind of limestone known as the cornbrash, from which arise 'brashy' or rubbly soils, is ideal for growing corn. The inferior oolite limestones form the great scarp of the Cotswolds and, as a finely worked building material, glow in the sun in countless villages in the counties of the Midlands. North of the Cotswolds the Northampton Sands are rich in iron-ore, and this once put in jeopardy from commercial open-cast mining the beautiful countryside around Swerford, Nether Worton and Great Tew. Much of England's heart with its landscapes and its houses reflects the long quiet period when shallow marine waters flowed over the land long after the great shudders in the period of earth-moving that raised the Pennines, Malverns and the Mendip Hills. Later, glacial deposits were to spread a mantle of drift material over the Midlands and these show evidence of having been brought from some distance. We are living in one of the quiet periods of the earth's long history and, as a naturalist exploring the English Midlands, I am constantly reminded of how the rocks and the emergent soils have shaped the vegetation and influenced the wildlife.

The rocks of the Jurassic period sweep in a narrow crescent from Dorset north-east to the coast of Yorkshire. The clays and limestones once supported damp oakwoods but a great deal of this has been replaced by pasture. It is, however, in a present Midland oakwood that I would like to begin my exploration of the rolling and very varied countryside on the borders of Northamptonshire, Warwick-shire and Oxfordshire. The wood can be found at about 500 feet above sea level partly on the upper lias and partly on the inferior oolite of the Northampton uplands. In April the disc-harrowed fields below lie golden-yellow like the sands of the Sahara but the hedges and ash-trees stand black and stark as in the very depth of winter. The oakwood folds itself over a ridge in the landscape. Tall standard oaks and some ashes, beeches, pines and larches tower over a secondary layer of spindly Spanish chestnuts, hazels and hollies, while the floor of the wood lies deep in last year's leaf litter. There are a few field maples, clumps of willow and many bramble brakes. Elders grow around an ancient badger sett deep in the wood, which in one part has been cleared and then replanted with conifers.

A favourite observation place of mine is just inside the wood, where birds tend to be more numerous than in the very heart. It is a dry, sheltered bank above a tiny stream which trickles its way down to join the River Nene in the valley. In spring golden celandines and silver-petalled windflowers and creamy primroses clothe the bank near the stream, hazel catkins toss and swing in the breeze above my head, and the green blades of bluebells yet to carpet the wood in shimmering blue push their spears upwards towards the sun. It was here that I often chose to listen to the dawn chorus of the birds. I remember one cool night with a slightly overcast sky. Just as I noticed the first lightening of the eastern horizon – an almost imperceptible change – some carrion crows woke up and, as they made off with raucous cawings, several woodpigeons began to coo moaningly in the distance. There was a sharp cold thrust to the wind but soon blackbirds, song thrushes and robins were singing. From one quiet glade – lit by a fitful rising sun – came the repeated notes of a chiff-chaff, not long arrived from his southern winter quarters. A blue tit 'seeed' away high up in the bare canopy of one of the oaks. Next to a sallow bush which was burdened with stout yellow catkins, where early bumblebees were at work foraging for nectar, I found a stump and here I rested quietly.

A blowfly settled on my boot and a tiny cloud of midges danced above the willow, shifting and gyrating like morning mist. A nuthatch whistled like a schoolboy. Yet the wood seemed partly asleep though ready to waken in an instant. I could hear the faint shouts of early rising children in the village below the brow and, near at hand, the comforting caws of rooks in their ancient suffering elms. A rank smell of fox drifted across the stump where I was sitting and soon I found his footprints in the muddy ride, running straight along a track and then finally disappearing into the long grass. The previous spring – just here – I had seen an old dog fox picking his way through the undergrowth. Two grey squirrels scampered down an oak tree to inspect the hunched human figure on the old stump. The dawn chorus – such as it was – was dying away and I set off to wander through the rest of the oakwood. In the region of young spruce and larch, then about twelve feet or so in height, I caught a glimpse of a small, foxy-brown creature disappearing into the conifers. Suddenly, it reappeared, leapt over a fallen branch and ran rapidly into the low vegetation with its head down. It was a muntjac – a small deer that stands only sixteen inches at the shoulder. The species was introduced to Woburn in Bedfordshire about 1900 and spread to Buckinghamshire, Hertfordshire, Northamptonshire, Oxfordshire and several other counties. Later I saw three others moving away from me, bearing bushy tails which were white underneath rather like naval pennants. I have walked through this wood for forty years and in the 1940s fallow deer browsed in the glades, but after a local shoot in 1948 only single specimens held on for a few years in the deepest cover in the wood. Now the shy and tiny muntjac has taken the place of the fallow deer. Other changes have taken place over the years as well – pheasants and tree sparrows have arrived in numbers as well as turtle doves and many more tree pipits.

Later in April the willow warblers, blackcaps, garden warblers, redstarts, tree pipits and cuckoos arrive. All three species of woodpecker nest, while tawny owls flit around and even hoot in the daytime. Moles burrow under the trees, stoats hunt along the rides, and the badgers bring out their cubs to play among the fallen but living elder boles, and to scamper and chitter and squeal over the fan-shaped piles of sandy earth at the sett entrances. As the tree canopy begins to grow and burgeon into green existence, the flowering plants on the floor of the wood reveal a distinct succession of bloom. Dog's mercury, celandine and bluebell give way to a progression of aestival flowers – wood vetch, tuberous pea, raspberry, brambles, bugle, veronica, orpine, sium, yellow pimpernel, yellow rattle and skullcap, while I know secret places for a clump or two of yellow archangel, a tiny patch of blinks in a muddy rut and two or three plants of broad-leaved heileborine orchid. The tall pinkish-purple spikes of rosebay, especially on disturbed ground, the delicate umbels of pink greater burnet saxifrage and the foamy cream clusters of meadowsweet bring an especial delight to the woodland in July. There are other rewards for the naturalist as well – grass snakes sun-bathing under the brambles, common lizards on the sandy outcrops, speckled wood butterflies dancing in the sunlit rides as well as small tortoiseshells, pearl-bordered fritillaries and in May brimstones and orange-tips.

For me a lowland oakwood in summer provides a richness of experience and a variety in its wildlife hardly matched by any other kind of habitat in the British Isles. Woodland has been called 'Nature's supreme development', and with its rich bird song, handsome trees, varied ground flora and the hum and glisten of myriads of insects' wings there is much to recommend it. The only drawbacks are perhaps the countless sweat flies that land on one's head and neck, not to bite but to drink up the perspiration – I use an elder branch above my head to keep them at bay – and the fiercely biting Tabanid females whose beautiful iridescent eyes belie their predatory nature.

As summer moves on, the badgers in the wood – now that their cubs come regularly above ground – begin some serious housework, throwing out old heaps of bedding and earth from the tunnels in the sett. Piles of freshly gathered oak leaves and bracken are stacked ready to dry and, when the adults are satisfied, each pile is brought in backwards, clamped firmly between the chin and fore legs of each animal. August nights are a rewarding time for watching deer, foxes and badgers and in a somewhat silent month for birds I look forward to the barks, yelps, moans and yarls – some of the thirteen separate sounds that badgers need to lead their lives to the full. How different are the oakwoods in winter with rutted, muddy rides while parties of tits, treecreepers and goldcrests wheeze and stutter among

'boughs which shake against the cold,
Bare ruin'd choirs, where late the sweet birds sang'.

20

Northamptonshire is also a land of fields, parks and lakes. Much of the region around the wood I have just described is grassland – ryegrass and agrostis. This provides food for many rooks, jackdaws and starlings which probe out the worms and leatherjackets. Skylarks and a few pairs of yellow wagtails nest as well, and swallows, house martins and swifts from the farms and villages sweep and call as they dive low over the grass collecting flying insects hatching out from the field below them. Small and large skipper butterflies are common in the grasses along the hedgerows, while meadow browns and gatekeepers 'wink' with indeterminate flight across the grassy swards. The rhythms of wildlife have become adapted to the slower tempo of summer as breeding comes to an end and the chirp of grasshoppers soothes us from the meadows. The yellowhammers still pour out their ditties from the hedgerows and the goldfinches whisper and flit like the thistledown on which they feed from seeding plant to seeding plant. Late summer means a heaviness about the countryside with the oily green of foliage forming a backcloth to the rough poppies, heartsease and corn marigolds which bow their heads in the dusty, almost suffocating summer wind.

Yet there is always the relief of water. A small pond near the edge of the wood is ablaze in summer with the flowering spikes of spotted orchid. One summer I sketched twelve separate flowers each showing a different pattern of lines and spots on the lip. Beyond the wood lies a piece of parkland called Fawsley once the home of the Knightley family where the Marprelate tracts were printed. It is a typically English park with a large house and a ruined Elizabethan dowerhouse of red brick, but among the grazing meadows, the cattle and the sheep are two large lakes. Below a fringing edge of alders, which attract siskins and redpolls in winter, are celery-leaved crowfoot, amphibious bistort, marsh hawksbeard and watercress, while the water itself is starred with white and yellow waterlilies. At the end of one of the lakes is a marsh where tall pokers of reed mace, the threatening leaves of arrowhead and reeds form a deep watery sanctuary in summer for reed buntings and many pairs of chattering reed warblers. Swans and mallard, coot and moorhens build their nests along the edge of the lake known as Big Waters, while little and great crested grebes – both gems of the waterside – breed despite the toll taken of their eggs by the carrion crows. As summer progresses, swifts come screaming over the lakes and as night falls Daubenton's bats join the avian hunters. In winter kingfishers come to the Big Waters, herons drop in and ducks such as teal, tufted duck, pochard and goldeneye join the mallard, coots and moorhens. Northamptonshire has other waters too – rivers, lakes and reservoirs, sewage farms which are all worth a visit. One marsh along the River Nene may hold up to 600 teal in winter with smaller numbers of mallard, shoveler and wigeon, which are often stirred up by a hunting short-eared owl beating its way over the grassy causeways where voles run among the docks and mushrooms. I have seen dunlin in tight little packs, straggles of lapwings like eyebrows in the sky, snipe with zig-zag flight and rarer ruffs and black-tailed godwits, and occasional green sandpipers. The sewage farms and reservoirs form valuable sanctuaries for wildfowl with Stanford and Pitsford reservoirs, which I often visited, even attracting goosanders.

The Grand Union Canal also crosses the heart of England and this I have explored by narrowboat from Braunston south to Wapping in London. There is something very restful and refreshing in a progress made at a sensible speed which allows one to watch the swallows dipping in the murky water and stare down at the shoals of tiny fish between the boat and the bank. I can understand the air stewardess on the South Africa flight who always insisted on spending her holidays on a narrowboat in England!

I remember with affection a pair of narrowboats that I joined at Braunston – the motor *Renfrew* and the butty boat *Lucy* that was towed behind. Perched on the blunt bow of the butty and listening to the steady 'chug-chug' of the diesel engine and the faint twitter of budgerigars inside the painted cabin with its brass lamps and lace plates I had a singular point of vantage. From here I could admire the engineering that had made such a feat as the canal possible. We climbed some thirty-five feet through six flights of locks just short of Braunston Tunnel and then refreshed ourselves with an alfresco pint and sandwich outside the 'Admiral Nelson' in the company of a rapacious Muscovy duck. From the uppermost lock the canal swept in a broad crescent towards the black hole of the tunnel – one and a quarter miles of horizontal shaft drilled through the hard ironstone of Braunston summit. With the headlight of an old Bentley car ablaze in the bow of the *Renfrew* we entered a dark dripping tube of mystery, emerging finally through a round hole into daylight and banks clothed with heavily-fruiting hawthorns, brambles, elders and nightshades. Flocks of starlings were feeding in the elders, squabbling over the harvest and then taking off in sprays of blue rain.

We paused at the smart Whilton locks, which were in the charge of a lock-keeper with a large white dog. Having once been fed regularly by a benign passing train driver, it had become obsessed with the electric trains on the Euston line nearby, following them as they roared by and even watching for the signals to change, in the despairing hope that another engine driver might throw him the reward of a sandwich. More than two hundred tortoiseshell butterflies were gathered on the magenta crowns of michaelmas daisies by the lock, seven swallows were sitting on a power line by the bridge and a linnet sang a thin autumn song from a thorn on the canal bank. Now the land level began to drop and we moved south to the rattle of paddle-like ratchets and the outrush of water. On the raised embankment above the honey-coloured church at Weedon two swans and some cygnets sailed past with a warning hiss or two from the cob. Just here one deep winter I saw an angler fishing through a hole in the ice, a paraffin stove between his legs! Just to the south of the long wet Blisworth Tunnel lies Stoke Bruerne with its Waterways Museum set among thatched houses, colourful boats and a white double-arched bridge reflected in the still canal water. The Grand Union meandered south to Tring with its feeder reservoirs, where I watched mallard, pochard, tufted duck, teal and a score of shoveler among the herons, coots and moorhens. Snipe, pipits and wagtails were feeding on the mud and I saw a single common sandpiper.

Not far from Tring tall serried ranks of dark green pyramidal Norway spruces, separated into compartments or blocks, and crisscrossed with tracks and rides, carpeted part of the hilly, rolling landscape. This was an alien scene quite out of keeping with the landscapes that I have just described and reflected the Forestry Commission's earlier determination to make good the losses in Britain's timber. On one of my visits to this State forest wreaths of white mist hung over the canopies of the trees. Under the dense shade of the trees carpets of needles made an acid ground where all vegetation was more or less suppressed. The sun's rays were able to slant between the upright poles of the taller trees, throwing pools of greenish light on the bare ground. Chaffinches were common and punctuated the morning with sharp 'spink-spinks' or their bursts of rollicking spring song. I could pick out wrens, goldcrests, coal tits and distant woodpigeons whose deep cooing songs were reflected from the wooded hillsides with a strange unnatural hollow quality. There were a few great tits foraging in the ash trees along the woodland rides and some firecrests hissed their high songs from the spruce canopies, very different from the commoner goldcrests with their trilled endings. Far above my head a pair of sparrowhawks circled and soared, and a crossbill 'jipped' loudly as he flew away from me. Despite the sameness of this plantation there were lots of rabbits, grey squirrels and muntjac deer and I spotted a weasel crossing a path at lightning speed. After this dog-leg into an alien

23

environment – an exception to the Midland rule – I must return to the heart of England that I know and love.

It is the nucleus of so many regions of Britain – the village – that reflects much of the countryside and its activities. There is the Northamptonshire village of Great Everdon, for example, with a main street of soft, iron-coloured houses and brick villas whose flowery gardens are full of bees and butterflies. Swallows and martins sweep past the four-square sandstone church and its graveyard where the forefathers of the village sleep. There are two inns – 'The Plough' and 'The Plume of Feathers' – as well as grassy swards, a manor and a little stream where the children paddle and catch fish. There are oakwoods nearby and all around lie rolling fields alternately green with pasture and gold with ripening corn. The summery lanes hum with insect wings, wasps have built in the thatch of a ruined cottage, and meadow cranesbill, creeping thistle, willowherb and hogweed flourish along their verges. In a cottage garden there are peacocks, tortoiseshell and red admiral butterflies and newly arrived silver-Y moths from southern Europe, each sporting its distinctive gamma mark. Below the village and alongside the cricket field are some allotments burgeoning with beans, peas, cabbages, carrots, onions and lettuces throughout the summer, but this cultivated patch which once seemed unlikely to be tilled when the present generation of allotment holders had passed on faces a new lease of life as food has become dearer.

Above the village lies Everdon Down; this was once the home of Neolithic man and I have picked up worked flakes of introduced flint in the cornfields and sheep pastures where the corn buntings' songs jangle like rusty keys from the hedgerows. By September bird sounds have become infrequent although the rooks and jackdaws, starlings and lapwings are flocking in the fields. Yet the bells of Everdon and, more distantly, those of Newnham ringing across the countryside often trigger off a late chiffchaff into song before he leaves for the south. There is also the musical twitter of swallows as they gather on the television aerials, perhaps on a cottage under the great east window of the church, from which the faint sounds of an organ proclaiming harvest festival mingle with the sweet voices of the birds.

In the evening pipistrelle bats emerge from the church tower to flit aerymouse-fashion around the rooftops of Everdon, Badby and Newnham. One April a bat attacked my father's Panama hat at three o'clock on a warm afternoon as he sat in his garden in Badby village – almost as unusual an encounter as that we had with a black cat in the main street of Newnham which fiercely attacked my father, myself and my dog and had to be beaten off with a walking stick!

All day long in late summer outside the village combine-harvesters move mammoth-like over the wheatfields, swallowing up the corn and discharging a cascade of golden grain as the rabbits explode in all directions across the stubble. Soon the stubble fields become roaring walls of flame, scorching the thorn hedges and the hedgerow trees as the stalks are set alight. Fire leaps across the stubble in terrifying fits and starts, while the blue, ash-laden air over the land dances and shimmers as columns of smoke tower up above the valley and plume away towards Northampton. I have seen many bush fires in East Africa but I find this scene in the heart of England alien and repugnant and in these days of advanced agricultural techniques without much justification.

Perhaps the most outstanding feature of this countryside is the amount of land laid out to grass, and here a hundred acres of grassland could support more than sixty sheep. Near the Warwickshire border and south-west of Daventry – now very much a growing conurbation – is the ancient earthwork crowning Arbury Hill which stands 735 feet above sea level – an inferior oolite top – with commanding views across the rolling patchwork of the English Midlands, and the striped pattern from the old open field system. From here I have watched many dawns rise, with the darkened countryside damp and impregnated with the strong smell of wet vegetation and sheep. In the distance cars' headlights have glowed sharply as the anglers came to park by Fawsley lakes with their large pike.

One October morning I watched a passage of small birds – mainly meadow pipits, skylarks, chaffinches, swallows, house martins and greenfinches – battling into a south-west wind. The valley of the Nene lay under angry, broken clouds – black and red – with clear golden strands. Down the broad flank of the hill a thin ghostly line of sheep wound its way, their complaining voices echoing hoarsely across the valley. They made their way down towards a small pond where frogs, toads and newts spawn in spring and from which I brought back plants and frog-spawn in a successful attempt to re-introduce frogs to my part of suburban north-west London. A moorhen was dabbling among the lily pads and a blue tit churred anxiously in one of the sallows by the pond. A magpie rattled away in the distance. The harsh notes gave warning of the advancing day.

Just a few miles west of Arbury Hill, across the Roman Fosseway, which like an arrow's flight links Stow-on-the-Wold with Wigston Parva on Watling Street, are 'Piping Pebworth, Dancing Marston' and a cluster of small Warwickshire villages west and north of Stratford-upon-Avon. Here green undulating country stretches wide

25

across to the scarp of the distant Cotswolds. There are thatched houses and cottages of red brick and timber on either side of the sedgy banks of the River Avon. I lived within a few yards of the seventy-foot-high maypole in Welford-on-Avon where on May morning the children from the village school danced and twined their gaily coloured ribbons. Around the house was an old-world garden of formal grass lawns, shrubberies, herbaceous borders, rosebeds and an orchard with rough grass, which all formed a small wildlife reserve in the heart of the village. A pair of rooks ejected from the noisy colony half a mile away built a nest of walnut and acacia twigs in the tall walnut that towered over the garden. 'Ah! I see you got a pair of crows in that there walnut,' pronounced one of the village sages. 'No! As a matter of fact they're rooks!' 'Rooks never nest by theirselves!' was the reply. It was impossible to argue with this rural logic but poor natural history.

Greenfinches, goldfinches and sometimes even linnets sang in the trees, chaffinches poured out their refrains from the orchard, and even a skylark soared in song over the garden as he drifted away from the fields close by. Whitethroats nested in a wilderness of nettles and raspberry canes and lesser whitethroats sang their rattling phrases from the tall hedges along the road; a pair of blue tits was nesting in the post-box across the street. A great spotted woodpecker used to drum on a dead branch of the acacia outside the bedroom window, and two tawny owls roosted in a tall Lombardy poplar. But Welford has changed over the years, with picture windows replacing leaded frames and bungalows spreading like some form of mushroom in the direction of Bidford. Yet I am glad to say that there are stretches of the river which winds its unhurried way past the old mill between rolling fields, creamy orchards and tree-tipped crests where rural England can still be seen in its most blissful and rewarding guise.

A lush growth of sedges spreads like a barrier across the stream where moorhens cluck and mallard call and little grebes – Shakespeare's 'die-dappers' – still dive. White clusters of deadly hemlock line the bank and masses of yellow water-lilies star the still surface of the stream. Sand martins hawk for gnats dancing in front of the arches of the old stone bridge, sedge warblers whistle and chatter in the willows and from a tall hawthorn come the fluent contralto notes of a blackcap. A water vole plops into the river, casting one last glance at the heron slowly flapping its way overhead. Far down the water meadows I can hear the lonely and infinitely stirring call of a curlew. A stone's throw from the bridge stands 'The Four Alls' – a reminder to the thirsty traveller that the countryman pays for all.

Across the fields from Welford and over the A46 lies a mixed farm of nearly 173 acres made up of thirteen fields of from seven to seventeen acres in size, an orchard and a small piece of woodland, lying alongside a brook. When I lived here I could look across to Meon Hill – the first bastion of the Cotswold Hills – crowned with a clump of trees on the ridge of the Roman encampment. There were Neolithic flints to be picked up in the fields as well as iron-brown fossil ammonites, polished belemnites like discarded bullets, and grey devil's toe-nails, relics of creatures that once lived in that same shallow Jurassic sea. Here I enjoyed a truly rural life with cows lowing outside the bedroom window, and our drinking water full of Epsom Salts – magnesium sulphate equivalent in power to Leamington Spa's Number 2 Strong and a trap for unacclimatized visitors!

I often walked round the farm with Crackers, my cross-bred terrier-sheepdog companion, who, although raised in the urban surroundings of Rugby where I had been teaching, soon became an expert rat-catcher and a delicate retriever of hens' eggs. One day he had his come-uppance. He flushed a barn owl which had been standing firmly clasping a dead rat. The startled bird flew up on to a fence-post and launched into an extraordinary dance, hopping from one foot to the other, but with its head held still and its gaze firmly fixed on the dog's face. Beneath the stare of those unrelenting black eyes the dog dropped down on to its belly and then slowly dragged itself backwards away from that pitiless countenance, and round the corner of a haystack. On another occasion, as I was sitting on a five-barred gate, listening to the reeling song of a grasshopper warbler, a dog fox emerged from a copse, and sat down only a few feet away; he saw the dog, tethered on his lead to the gate-post, but failed to notice me, quiet and above the line of his sight. One day the dog explored a badger sett in a field near a stream, disappearing for five minutes down one entrance and returning, somewhat sheepishly but to my intense relief, from another hole thirty feet away.

Altogether I observed nearly ninety different species of bird on this farm. One year a whole field was given over to sunflowers and brought in bullfinches and greenfinches from miles around. Rooks nested in the trees by the Marchfont Brook and roosted near Meon Hill. Reed buntings bred by the farm pond and a pair of tree pipits had their territory on the road to Quinton. The male pipit would flutter up at a steep angle from an ash tree, start to sing his opening notes and trills and float down with wings bent upwards and tail fanned, giving out a series of 'seeea-seeea' notes, to land once again on his favourite branch. Seven different kinds of warbler nested in the copses, and there were twenty-four different species in the orchard, including magpie, green and lesser spotted woodpeckers, treecreepers, redstarts and goldfinches, and a pair of little owls, while a visting sparrowhawk often hunted through it as well.

An orchard with mature trees which remains unsprayed can be rich in arboreal birds that bring us all delight; those that are newer and managed in the modern manner can be very much poorer. There were also farmland birds such as pheasants, common and red-legged partridges, lapwings, barn owls, skylarks, cuckoos, tree sparrows and yellowhammers. Rarities included passage Greenland wheatears, whinchats and stonechats, peregrines, merlins and a rough-legged buzzard, but redwings and fieldfares were common and white-fronted geese flew over in December.

The grassy roadsides were rich in summer flowers – blue meadow cranesbill, royal purple knapweed and yellow ladies bedstraw. Butterflies flew over the spikes of grass and heavy umbels of the hogweed – browns, skippers, blues and magnificent marbled whites whose caterpillars could be found feeding on various grasses on the roadside verge. I often saw stoats and weasels hunting these verges and entering the dense woodland scrub at nearby Preston Bushes, where about the 26th of March I used to hear the first chiffchaff of the year. About 16th April the nightingales came back to the scrub to sing until the middle of June, gracing the day more than the night with their notes. Other nightingales used to come to the thorn bushes along the old tramway, whose route can still be traced and which was designed early in the nineteenth century as a horse-propelled railway to link the Worcester and Birmingham Canal and Avon navigation in Stratford with Shipston-on-Stour, but which was scrapped in 1881.

The Warwickshire roadsides were often left uncut and unsprayed and were rewarding areas of exploration for naturalists. Among the invertebrates were carnivorous centipedes, which preyed on the slugs and insects that shared their habitat, grasshoppers that chirped in the warm sun, woodlice and nocturnal earwigs hidden under rotting branches, hoppers that spread their foaming masses of white spittle on the roadside plants, while scorpion flies and hoverflies and black-tipped soldier beetles alighted on the domes of hogweed flower. Occasionally a large hawker dragonfly from the pond on the farm would come and quarter the miniature jungle by the road.

Preston Bushes were the home of speckled wood butterflies and here I also filmed a spider – *Pisaura mirabilis*, a common animal in Southern Warwickshire – and its nursery tent of silk made to protect her round egg-sac as the young spiders begin to hatch. The young appear in late June or early July and, after an initial moult, which takes place inside the tent, gather in a tight ball for a few days before scattering from their silken home. Here is a riot of early purple orchids, red campion and ground ivy – a natural garden for plants, mammals, birds and insects that rejoice in open and sunny thickets. In July they are also alive with shrews, whose high-pitched squeaking may sound like the calls of birds but come instead from the diminutive insectivores with faces like those of little pigs. This region of South Warwickshire in autumn sees in some years a remarkable passage of birds – a phenomenon that I first discovered in 1948. This represented one section of an inland migration route from the Wash to the Severn and involved thrushes, pipits, larks, wagtails, swallows and house martins, finches and two species of gull. Above the farm and along the steep scarp of the Cotswolds in the early mornings of September and October the watcher may observe a fascinating passage of migrant birds. I would often, with John Drinkwater,

> see the valleys in their morning mist
> Wreathed under limpid hills in moving light

and then an hour or so after dawn see birds moving south-west from the farm to Meon Hill and then on to Cheltenham. I recorded in all nineteen species on this flyway, and on one September morning counted as many as 1,167 birds going through; such a morning was indeed a concrete revelation of one of nature's recurring marvels which I had also glimpsed over Rugby, Windmill Hill, Stanway, Birdlip and further down along the Cotswolds. One September day as I was walking along the ancient earthworks on Meon Hill watching migrants I met a tall stranger who spoke with a slightly Yorkshire tongue. He was thrilled and fascinated with the countryside and quite overwhelmed by the view from the top. Then he spoke of duck hawks and sharp-shinned hawks and vertical rather than horizontal bird migrations and I knew he was from the New World. He was originally from Yorkshire but his home for many years had been Tucson in Arizona. He had not been home to England for fourteen years. I would not have missed this conversation with a man who was seeing our countryside with eyes accustomed to desert plains and basins. Like all those who have been away for a long time he was noticing details that might have escaped the ordinary observer. Standing high above the Avon plain it was almost possible to imagine the Roman eagle on the citadel of the camp, staring south-west along the Cotswold Hills.

Much has been written about the Cotswolds and it is not my intention to go over old ground. Yet there is a hidden inescapable beauty about many of the villages nestling in the folds of the high oolite wolds that unroll as far as the eye can see. In the west and north it is a land of enchanting beechwoods whose flora holds such interesting plants as common wintergreen and red helleborine – one of the shyest of English orchids. Over the northern, eastern and central wolds the oolite grasslands are grazed by sheep or ploughed; here the naturalist will find torgrass, meadow cranesbill, mallow and scabious and especially fine specimens of musk and frog orchids. A concentration of limestone shapes the scenery and fashions the flora and the fauna. The buildings and walls are honey-coloured and gather up the light, standing embosomed in a delicate landscape which is criss-crossed with clear rivers and lesser rivulets flowing towards Avon, Thames, or Severn and punctuated with the exclamation marks of the upright wool churches. Houses and churches declare their own style, formed with a delicately responsive stone and enriched with gables and thin stone roofing tiles, all revealing, according to Alec Clifton-Taylor, 'the visual accord which they achieve with the landscape in which they are placed'.

For me who lived and worked in the Cotswolds there are many special visual memories. I recall the especial pleasure of the grey wagtails flying upstream as I drove through the water to Duntisbourne Leer, the pied wagtails flitting through Ozleworth Bottom, larks weaving a silver chain of song above Amberley and high Snowshill, a kestrel hunting in the Golden Valley above Sapperton, martins dipping into the stream at Lower Slaughter and swifts screaming in front of the gold and grey houses of Stanton. There was the mysterious long barrow at Belas Knap, whose short turf was starred with aromatic thyme and rock roses and where the butterflies and bumblebees flew just as they must have done when the dead were buried in this mound with its drystone walling, perhaps four thousand years ago. In September sunlight I have also stood in Bibury and marvelled at the stone cottages and their gardens, the roses in the churchyard, the Tudor manorhouse and the young children dancing in the open space near the school, all of which combined to imbue that village with a sweet enchantment. Scores of fish swam lazily in the river below Arlington Row and a late chiffchaff sang a few phrases from a willow.

Near Tayton the river Windrush, bright with monkeyflower and Himalayan balsam, wound its way beneath the higher land where the Forest of Wychwood once stood. Disused watermills stood empty and silent along its course and in the pools coots and moorhens dredged up weed. How different this scene was from a spring marsh on another Cotswold river – a veritable wilderness of tussocky grass and rushes set about with tiny alder trees – where I saw the lapwings rising and tumbling in their spring ecstasies, snipe which, driven by the spring fever in their blood, circled and dived in their strange bleating display flights, and redshanks – the 'wardens of the marshes' – yodelled in their excitement above the swampy ground.

Not all the landscape in this heart of England is open to the winds. There is Great Tew, for example, near Chipping Norton whose manor was replanted in the nineteenth century by John Loudon. His evergreens add a touch of mystery and awe to this changing hamlet. In autumn the collared doves add their own note of melancholy to the long laurel walk that leads to the church, and grey squirrels whine and fuss in the trees beyond the churchyard. The village street hangs heavy with thorns and yews whose berry harvest brings in the thrushes from the countryside around. There is the strange tapping sound too as a song thrush uses the roadway as an anvil on which to smash its snail shells. Not far away is the ancient stone circle of the Rollright Stones – smaller than Avebury and much weathered on its ridge. Here there is a strange air of gloom and foreboding, relieved only by the corn poppies and the casual phrase of a passing linnet. How could there be plans for a caravan leisure site close to this ancient place of mystery? In the words of Robert Burns with 'weary winter comin' fast' the fields turn into mud lying thick and liquid under the leafless trees. The countryside takes on an air of decay and the farmworkers warm themselves by the fire that denotes another fifty yards of hedge-laying accomplished.

To Hilaire Belloc the Midlands were 'sodden and unkind' but I suspect he was thinking not only of the countryside! For me their fascination and attraction lie in the way in which they reflect the changing seasons. March gales may bring the beeches toppling down but the skies can be of a compelling cerulean blue. Snow may lie along the bare headlands of the fields but soon the hedgerows and copses will be alive with bird songs, the hum of insects' wings and the drift of English flowers. It is in the little series of vignettes that I have drawn in this chapter that the reader will I hope glimpse something of the variety of the English Midlands and yet see something of the underlying common strands that make the heart of England what it is.

2. Western Moors

The Cotswold Hills, with which we were concerned in the last chapter, come to an abrupt end east of Bristol at Wootton-under-Edge and Stinchcombe Hill. From here the eye is free to wander across the Vale of Berkeley to the Severn, while to the south-west lie further ranges of hills – the Mendips, Poldens, Quantocks and beyond them Exmoor and Dartmoor. Westward is a higher land of sedgy moors which has a wildness and touch of mystery different from the more comfortable woods, fields and quickset hedgerows of the English Midlands. There are three chief folds in the landscape that give the county of Somerset its variety and appeal. The Mendips – a mass of carboniferous limestone with caves, gorges and dry valleys – roll on from Frome to Weston-super-Mare and then push up outposts in the Bristol Channel – the islets of Flat Holm and Steep Holm. Beyond are the moors and swamps of Avalon and Sedgemoor, rich in Arthurian legend and dominated by Glastonbury Tor, the home of nightjars, marsh and grasshopper warblers. Across this countryside stretch the low, graceful Polden Hills of blue lias with their holy wells and often dramatic views of the Mendips brooding over the peat moors from a damp grey sky.

Where the Somerset Levels reach the coast many salt marshes and flats have formed, and here under the shadow of Hinkley Point atomic power station many wildfowl and waders congregate and feed in peace. North of Taunton the Quantocks and to the south the Blackdown Hills are geographically and geologically detached parts of Exmoor and both are formed from tough Devonian rocks. Dartmoor to the south is a stupendous mass of granite capped with great sweeps of bracken and heather-covered moorland and ornamented with conspicuous tors; it is linked to the igneous outcrops that form

Bodmin Moor, the Land's End peninsula and the Isles of Scilly. The West Country is a favourite region for holidays, and this is largely due to the granite uplands and exposed rocks, the red cliffs, the wooded combes and sheltered lanes that by their variety and different appeals can so readily waylay the traveller. In this chapter I hope to reflect the character as well as the wildlife of these regions, and especially those to be found in the higher lands of Mendip, Quantock, Exmoor and Dartmoor, which I have chosen to call the Western Moors.

The Mendips extend as a steep-sided ridge of hard, crystalline limestone. North of the cathedral city of Wells the ridge is about six miles wide and rises to over a thousand feet where the land is acid. The ridge was uplifted at the end of the Carboniferous and the start of the Permian period. There are many features of outstanding interest for the naturalist – a tableland that can be bleak and windy, vast cracks and crevices brought about by the dissolving of the limestone by water, whose sides support many unusual lime-loving plants, and the remains of old mines worked as long ago as Roman times. Of the many beauty spots in the Mendip Hills Cheddar Gorge is without doubt the best known. In summer it is full of coaches and tourists and I prefer to visit it, say, in March or early summer when the Gorge is empty and almost entirely one's own.

One spring day I drove across the plateau, pausing near Priddy's ancient barrows to listen to a sweet chorus from some thirty redwings gathered in a small copse. I explored the crumbling walls and peered into the shafts of the old Roman lead mines, reminiscent of those I knew in the Pennines. Legend has it in this rather desolate region that Christ passed this way as he travelled through Mendip. The limestone itself has been used to create many beautiful buildings. My hill road began to drop in curves and twists down into Cheddar Gorge to the accompaniment of numerous skylarks singing above the ridge, while a dark brown buzzard, frightened by the car, rose up and on finding a thermal began to climb higher into the sky.

The Gorge itself is dramatic – a high towering ravine which had probably once been a roofed chain of caverns whose natural vault had at some time fallen in on itself. The sides of the Gorge are rugged, almost vertical rock-faces rising to more than four hundred feet. Many of the ledges and cracks provide footholds for small dark yew trees, paler ash trees and ivy which falls in cascades down the cliff-faces. From the towering walls of limestone were reflected the harsh calls of jackdaws, the high trills of several wrens and the chatter of courting kestrels, which circled and pursued each other through the ravine. To see some of the floral gems it is necessary to

33

visit the Gorge in summer, braving the lines of cars and coaches. For this reason I go there in the early morning so that besides the bird songs to delight me I can still admire the drifts of valerian, wallflower, oxeye daisy or Welsh poppy, rock stonecrop, mossy saxifrage, lesser meadow rue and fortunately for them, high up on the cliffs, the Cheddar pinks. As the sun rises, so the bees and two-winged flies add their gentle hum to the sounds of humans stirring in the village nearby.

Away from the Gorge there are other flowers to delight the naturalist – cut-leaved selfheal, blue gromwell and the low greyish umbellifer, the rare honewort, that flowers in May and June. Brean Down on the coast, which I have reached by travelling in between the dense sea buckthorn thickets around Burnham, and from which the last Roman soldiers are said to have left in AD 411, is still an important location in Britain for the white rock rose.

In Britain we owe to the carboniferous limestone a great deal of the beauty not only of Mendip but also that of Malham Cove, Dovedale and the Great Orme. The Mendip Hills are riddled with a labyrinth of underground water courses and cave formations. Commercialized though it is the stalagmites and stalactites in Cheddar in Cox's and Gough's Caves still provide a wonder for the eyes. Wookey Hole near Wells consists of a series of huge underground chambers made by the River Axe, which rises several miles away in a pool near the village of Priddy. From here a stream overflows and then after a few yards disappears into a rocky hole decorated with hart's-tongue ferns – a swallet – to be lost from view. One August I stayed in the village of Wookey itself a few miles from Wells, with its swans and grey wagtails. Early morning was grey and still and the clouds hung low and heavy on the Mendips. Just before half-past-five parties of noisy rooks with a few jackdaws began to gather in the trees across the street. A wren began to trill just before six o'clock; this was followed by the cooing of woodpigeons and the chatter of starlings from the television aerials. A swallow chirruped on a telegraph wire in the main street and for a minute or so its trilling dominated the bird sounds. House martins began to hunt for insects about the rooftops and I was able to tape-record their dry twitters and the soft sweet song which is often associated with dawn.

But I was just as interested in natural sounds *below* as those above the ground. Through the kindness of the late Wing Commander G. W. Hodgkinson I spent eight hours down in Wookey Hole. To reach the cave system I left the car park close to the building that then housed that famous dolls' house – Titania's Palace begun in 1907 by a doting father – past one of the oldest paper mills in Britain and along a wooded walk. The entrance tunnel above the escape route of the River Axe was inhabited by Celtic peoples from the Continent from 250 BC to AD 450. The first great chamber of the cave adorned by the stalagmite known for a thousand years as the Witch of Wookey is the Witch's Kitchen; here the River Axe flows blue-green and quite fast through the cavern. Here in the darkness and the middle of the night I was able to listen quite alone to the strange, eerie drip of water falling from the chamber roof more than a hundred feet high. Tunnel roofs were pitted with depressions carved by water, that in the classic illusion switched imperceptibly from hollow to bump and back again to hollow. I entered the third chamber with its span of 135 feet and maximum height of only twelve feet – a flat arch which would seem an impossible piece of engineering! At the far floodlit end the river enters through a completely submerged arch – the only way for divers to reach the inner chambers. Beyond the explored area may be mysteries still unknown – quiet pools and rapids, cataracts and falls – while the Axe may flow peaceful and sluggish one day and a roaring rushing torrent the next.

An offshoot of the great cave of Wookey is the Hyaena Den, which lies on the east side of the valley and where I heard wrens and hedgesparrows in song. This cave was found to be full of the remains of mammoths, woolly rhinoceroses, cave lions, cave bears, sabre-toothed tigers, wolves, bison and reindeer as well as flint weapons and implements. The den was alternately inhabited by men and hyaenas some sixty thousand years ago. The hyaenas hunted their prey animals over the 200-foot high cliffs which are now covered in dense creepers, and tower above the gentle cascade where the Axe escapes from its underground passage. Outside, the day was one of hot sun, thunder and heavy rain, but in the dark fastnesses of Wookey Hole, time and weather had no meaning and even space seemed to take on another dimension. Yet with the rain in a few hours this too would change with fresh surface water pouring through the caverns.

The Mendips remind me of the Derbyshire Peak but here the limestone lies over the red sandstone. Once mined for lead by the Romans and for coal around Radstock and Midsomer Norton at least since the fourteenth century, and also hunted over for deer by the Norman kings, the concerns of the present inhabitants are still quarrying and farming. The Wednesday closest to the 21st of August is traditionally Priddy's annual sheep fair. In summer the high land is the home of corn buntings, wheatears, and ravens and I have occasionally heard woodlarks in song. A grasshopper warbler was reeling away in Burrington Combe where in the eighteenth century

35

the Reverend Augustus Toplady, sheltering from a storm, was inspired to conceive the hymn 'Rock of Ages'. It is not far to Sedgemoor, with its pollarded willows and red-brick farmhouses. East from Burrington the road slips through dense woodland, skirts a lake and cascade and drops down to Blagdon; here there is a great reservoir a mile long and with a surface area of some 450 acres. I have seen this man-made lake at many different times of the year – in winter with its divers, countless duck, swans, geese and water pipits, at the seasons of migration with its ruffs, curlew, sandpipers, dunlin and terns and in summer when the great crested and little grebes and the garganey are nesting. I have a particularly clear memory of one March day when I stood on the road that runs along the embankment and listened to the goldcrests in a cypress tree singing their little songs with terminal flourishes. A common sandpiper, probably an over-wintering bird rather than a recently arrived migrant from Africa, was feeding at the water's edge. Near the embankment was a great overspill from the reservoir, and on this wet ramp I could see many chaffinches and meadow pipits as well as thirty-four pied wagtails, six grey wagtails and an adult male white wagtail with a pale grey back and rump – an early migrant. The grey wagtails were busily flitting back and forth across the road and giving their ringing metallic double calls.

From Blagdon I have usually moved on to visit the newer reservoir at Chew Magna. This is at least three miles long and lies in the flooded valley of the River Chew to the north of West Harptree, once the centre of the teasel industry. Chew Valley Lake is a truly idyllic spot, set against the backcloth of the Mendips and in a landscape of patchwork fields and woods that descend gently to the borders of the lake. In spring I have watched hundreds of coot and duck sheltering from the wind in tiny bays along the shore. There are often many teal and tufted duck as well as wigeon to be seen quite close to the road from Chew Magna. Pochard are there as well, in the company of little grebes, and one of the other winter delights is to see drake pintails with chocolate-brown heads and necks and needle-points of tails. Mallard are very common and shoveler and teal can regularly be seen, and I have sometimes spotted a rarer goldeneye. Up to five thousand ducks may occur, representing well over a dozen species. The basin of the lake often resounds with the loud boisterous quacks of duck mallard and the soft whistles and 'yeebs' of the drakes, the deep penetrating coughs of gadwall, the lower grunts of shoveler, the high musical notes of teal and the melodious slurred whistles of the drake wigeon and the rolling calls of the ducks. There may be rarities too like the pied-billed grebe from America, which frequented Blagdon in May 1968 and actually 'sang', while ring-necked ducks have also been also seen on these western reservoirs; my son David saw three one winter.

The Quantock Hills to the west beyond Bridgewater are in effect outliers of Exmoor. This ridge of hills about twelve miles long and three miles wide rises to a height of over twelve hundred feet. In some ways it resembles Exmoor but it is less bare despite areas of bracken, heather and gorse, and is much more heavily wooded with tree-lined valleys and coombs to give it a softer appearance. The scenery is milder just as the landscape of Exmoor is less rugged and awesome than that of Dartmoor with its stark, bare tors. In parts of the Quantocks I find a suggestion of the Midlands but it is more exaggerated here with pronounced whale-backed ridges of sandstone, broken by deep coombs and laced with hedges of beech. There are also many fine stands of beech, where I have heard wood warblers giving their shivering trills just as they delighted Gilbert White in the beeches of Selborne Hanger. There are many quiet winding lanes and above Alfoxton Park, where Wordsworth once lived and talked with Coleridge, is a breathtaking view across Bridgewater Bay to the Welsh Mountains. The chief valleys in the north are Holford and Hoddens Coombs, clothed with oaks and starred in spring with windflowers. There are a few deer in the Quantocks as well as ponies, foxes and badgers, while woodpeckers, redstarts, tree pipits and warblers are regular in the coombs, and I have glimpsed a buzzard flapping out of a tree on several occasions.

As one climbs up through the coombs between banks of golden gorse and later purple heather, where I have seen stonechats and whinchats, it is worthwhile to pause and look out for grey wagtails flicking their tails up and down, and for sprightly pied dippers on the streams. Across the barer summits with their whortleberries and ling a raven may pass, 'kronking' in alarm. In 1963 there were red grouse near Robin Upright's Hill and there used to be blackcock but there are no recent confirmed records of breeding. At the northern end near East Quantoxhead cirl buntings can be heard stuttering out their short rattling refrains; they are really birds of the Mediterranean area and in south-west England favour coastal situations and well-wooded valleys. They have been decreasing perhaps through loss of habitat and the disturbance that arises from new housing estates.

'Exmoor for beauty, Dartmoor for wildness!' The plateau of moorland that is Exmoor, with its National Park, has in some places changed little since the days of prehistoric man. Its folds glow warm and rich in summer while winter frost and blizzard can change it into

37

a white wilderness where the deer, ponies, cattle and sheep struggle for their very lives. There are wild untamed heaths riven by secret coombs of oak and thorn or deep small valleys noisy with waterfalls and rapids like the famous Doone Valley. The wooded Horner Water, shadowed by the 1700-foot peak of Dunkery Beacon, is one of my favourite places. It lies between Porlock and Luccombe and in spring there is a special beauty about the alders and the birch trees, the dark oaks and paler ashes. These are the woodland homes of the summer warblers that we met in the heart of England. Chiffchaffs, willow and wood warblers, garden warblers, blackcaps and white-throats join redstarts and tree pipits and the resident tits and thrushes in a great chorus of sound. I have seen nuthatches and lesser spotted woodpeckers going to their nest-holes and have listened to the sharp 'quet-quet' of great spotted woodpeckers as I walked up the coomb. It is also worth listening for the opening 'Syet-syet' or 'Per-fu . . . per-fu' of the handsome black-and-white pied flycatcher, which introduces a cadence or arpeggio of pleasing liquid notes. Tarr Steps is another place where it often sings. There is a good representation of woodland birds in the Exmoor coombs. The redpoll's wheezing trill is another bird sound that I heard above these woods. It is also a fine sight to watch a buzzard soaring and banking over the trees or a sparrowhawk dashing past the boles of the oaks. The summer bird songs drift on breezes rich with the scents of rowans, honeysuckle and ground ivy.

There are dormice in these woods and badgers, as well as hedgehogs and hunting stoats and weasels. I have seen plenty of rabbits and a few red squirrels among the oaks of Horner and elsewhere on north-eastern Exmoor. It was E. W. Hendy who, after watching a squirrel in Horner Coomb, wrote that it was 'almost bird-like in his constant expenditure of nervous energy and *joie de vivre*'. Exmoor sometimes undergoes appallingly wet weather and these wooded coombs can be dank and their misty air redolent with the slightly musky scent of sodden undergrowth and trees. Exmoor, unlike the Quantock Hills, which have been designated an area of outstanding natural beauty, was confirmed in 1954 as a National Park covering more than 260 square miles of wooded ravine, moorland and coast.

As you climb up to the wilder parts of Exmoor the natural history changes. I know one coomb with wild contorted sides where stunted beeches and twisted oaks have their last perches before giving way to a more open region with dwarf rowans. Here in summer the blackbirds in the woodland fragments also give way to their migrant cousins – the ring ouzels – whose plaintive wild songs come floating from a perch on an old stone wall or miniature thorn. The ring ouzels nest in the coomb often near a rock where the heather provides an overhang. This is a remote rewarding place for the naturalist, with the clear bubbling waters of its stream, its heather, bracken and scattered rowans. Here too the merlins hunt, sweeping sickle-winged across the moor and nesting perhaps in an old crow's or magpie's nest in a stunted rowan. In the days of falconry the small stature of the merlin reserved it as the 'Ladies' Hawk'. With its dashing flight and attractive plumage it is one of Exmoor's ornithological gems. Most nests are just scrapes in the bracken or heather and the young are pugnacious and restless, with an insatiable appetite for plucked small birds. The males generally have special plucking places – a boulder or a mound away from the nest – and several that I have come across have been fringed with the scattered feathers of meadow pipits and stonechats. The fierce little merlins can be seen swooping through the coomb to the alarm of the small birds. And this is very much a region of small birds – pipits, chats, the wheatear, stonechat and whinchat, and wrens. A glance up at the sky might reveal a mewing buzzard or a honking raven adding to the wildness of the scene.

Many ferns grow along the coomb sides and foxes have their earths among the fronds. I have seen a line of red deer making its way slowly across a rise, to stand for a minute or so silhouetted against the sky. On the slopes of Dunkery Beacon the red deer roam wild as they have done since prehistoric times, largely because parts of Exmoor are not accessible to cars. In fact, I once came down a narrow road off Dunkery in a car to find a notice at the bottom saying that the road was unsuitable for motor vehicles! Exmoor was once a Royal Forest and, although disafforested between 1815 and 1819, it is said that staghounds were kennelled at Simonsbath as long ago as the sixteenth century. It is not for me to enter here into all the arguments about stag hunting, although personally my heart reacts against the apparent infliction of pain or fear on an animal for our own gratification, and any civilized person must surely question the propriety of a stag being brought to bay in a school playground. Deer do have to be managed for their own survival and in excess they can cause damage; elsewhere they do seem to be quite properly conserved without the aid of hunting. Nevertheless, the red deer of Exmoor are a feature of the scene just as are the lizards that glide over the sandy red soil and the adders that bask on the warm earth. In October the coombs resound with the belling roars of the stags as the beeches are turning to bronze, the birches begin to take on an ancient golden sheen, the leaves of the brambles and dogwood assume a reddish hue and there are luscious clusters of purple, black and scarlet berries in the hedgerows. Clumps of red campion and yellow toadflax still flourish in the misty gossamer-laden air of autumn. The top of Dunkery Beacon stands hard against a sky full of golden light, laced with strands of pale pink and green and the amber hunter's moon shines across the valley as the roar of the red deer stags floats out of Horner Coomb and, farther away, from below Robin How.

Exmoor's wildest parts lie either side of the River Barle, whose ancient packhorse bridge with its seventeen spans of stone slabs is a mecca for visitors to the moor. Not far from Tarr Steps is Winsford Hill, where I have seen some of the moor's red grouse. Attempts were made in the early 1820s to introduce these gamebirds to Exmoor but these failed and the present population, like that on Dartmoor, is the result of further introductions during the First World War. They can also be seen on Dunkery Ridge and Withypool Hill. The much rarer black grouse can also be found on Exmoor, but this isolated community of birds has been reinforced by hand-reared birds that were released in 1969 and 1971. One Exmoor 'lek', or assembly ground, where the black grouse displayed held up to twenty-seven males and seventeen females; this was in the early 1960s near Chetsford Water. The hisses and roo-kooing of the black grouse on a May morning would mingle with the persistent songs of the pipits and the calling of the cuckoos in the coomb below. There is a great deal to interest the naturalist on Exmoor. Some hillsides are graced in spring with the sad sweet notes of woodlarks and in summer by the steady spinning of nightjars. There are also the sea coombs where water courses flow down towards the waters of the Bristol Channel. These 'goyals', as they are often called, may be treeless – just turf and rock where one again meets the wheatears, stonechats and whinchats of the more inland valleys and the adders, badgers, foxes, stoats and weasels. Buzzards may nest on the cliffs where the coomb reaches the sea. Other sea goyals are filled with stone screes and dwarf oaks or rowans, and for those who love true solitude they are some of the most remote and pleasing parts of the British countryside.

Dunkery Beacon overlooks Porlock lying at the head of a beautiful vale. Porlock commands attention, with its harbour, fifteenth-century Doveray Court and thirteenth-century oak-shingled church and its whitewashed cottages, its associations with Samuel Taylor Coleridge's Xanadu and Robert Southey's stop at the Ship Inn, where he found himself 'by the unwelcome summer rain confined'. Porlock Bay with its pebble beach lies below the Weir and Bossington where the nightjars whirr away in the late summer. In these days the shingle is gilded with the bushy felt-covered stems of silver ragwort – a naturalized composite flower from the Mediterranean. I have made my way through the ragwort to the marsh which lies behind the shingle beach. This is a region of pools, drains, mud, small reedbeds and rough grassland centred around an old duck decoy. In summer it is the haunt of coot, moorhens, many sedge warblers and several pairs of reed warblers and redshank. In the seasons of migration the marsh attracts waders and ducks, especially teal, shelduck and sometimes pintail. When I have walked across the marsh in September I have come across bar-tailed godwits and greenshanks, while whimbrels, ruffs, little stints and sandpipers drop in for a short time. At this time the blue-grey of the hills and the pastel shades of west country skies are reflected in the still waters, from which a curlew has just risen and whose haunting calls travel over the sedgy wilderness.

For two Septembers I lived on the edge of Dunster Beach beneath the shadow of Dunkery Beacon and not far from the village of the same name, with its castle dating back to the eleventh century, its navigational Conygar Tower, its ancient Yarn Market, old cottages and medieval Hobby Horse dance. The coastline here stretches north-west to south-east and from my base I could look across the Bristol Channel to the hills and conical tips of South Wales, the twin humps of Flat Holm and Steep Holm, and the long limestone bluff of Brean Down.

Behind the beach is a long narrow freshwater lake with sallows and a bed of reedmace where up to five hundred starlings roosted. Each evening before dusk I used to watch parties of starlings sweeping in from Exmoor and performing the most intricate and wonderful displays in the air before perching for a short time in some Scots pines and then plunging with loud shrieks into the closely packed reedmace. The lake was also used by parties of mallard, which often flew out to sea to join smaller numbers of shelduck, scoter and newly arrived migrant wigeon. Migrant warblers and nomadic parties of tits explored the sallows for food. The most noticeable of the warblers were the chiffchaffs, which sang regularly and the willow warblers which could hardly manage a note. A flowering buddleia near the back door was a mass of peacock and painted lady butterflies while on the beach nearby white wagtails hunted flies among the pebbles. Wheatears would often sit among the Duke of Argyll's tea-trees on the West Somerset golf course or on the clumps of fennel, where I once saw a little bunting feeding.

Each day I used to explore the edge of the golf course. One morning I walked among some forty-eight tired golden plover that had flown in overnight and on another day I counted 480! The

41

greens were favourite feeding places for waders– flocks of up to 350 curlew and 200 lapwing as well as a hundred oystercatchers, turnstones and smaller numbers of dunlin, ringed plover, redshank, godwits, ruffs and even nine sanderlings – those tiny beach-loving wading birds that run back and forth in front of the waves like clockwork toys. These would remain until the greenkeepers arrived to tend the turf and sweep it clean. There was also a hen merlin down from Exmoor which used to hunt the frightened dunlin out over the sea.

I had records of spotted redshank and wood sandpiper but perhaps the most interesting aspect was the series of coastal migratory movements that took place on my very doorstep. Swallows, skylarks and meadow pipits could be seen flying to the east at Dunster but to the west at Porlock. I regularly saw these species flying over from the Welsh coast and on reaching Minehead divide into two columns moving in different directions. On many days there were whimbrel flying west as well as kittiwakes and common gulls. On 2nd September 1963 I actually counted 660 black-headed gulls travelling north-west. There were smaller numbers of Sandwich and little terns and single little and Sabine's gulls; the rare Sabine's gull from the Arctic was being heavily mobbed by many of the black-headed gulls. Here on the edge of Exmoor I was able to watch birds on the move, to find peace and a refreshment of spirit and an opportunity to re-examine and reassess contemporary values.

Whilst Exmoor is a land of sandstone and shale, Dartmoor is one of the Armorican granite masses, igneous in origin. Three hundred million years ago molten rock was forced up into the slates and sandstones already present; then the rock cover was eroded away and the granite dome remains very resistant to weathering. Here is the largest area of windswept moorland in the south-west of England, covering some 240 square miles. There are summits of 1,500 feet or more on Dartmoor. At these high altitudes the damp air and heavy rainfall produce regions of wet moor and blanket bog similar to those in the Scottish Highlands or on the Pennines, but without specifically alpine plants. There are long, wide views across gently rolling rough grasslands and bogs, bracken- and heather-covered slopes towards distant skylines on which rest blocky outcrops or 'tors', which remind me of the inselbergs or kopjes that I saw in East Africa, fashioned by chemical and thermal wear. Some tors resemble ruined castles or forts while others seem just random assemblages of rocks riven by both vertical and horizontal joints.

Dartmoor is a land of mist and moor, imposing rock outcrops, a National Park, where on a fine day of sun and blue sky there is nowhere more benign, yet it is also a region of harsh stone, treacherous green mires and when wet or cold is as lonely and forbidding a place as anywhere in Britain. It demands respect at all times, a compass and an Ordnance Survey Map. I particularly remember walks towards Hay Tor, Rough Tor and Kitty Tor, where the sun has been warm on me and the air has resounded with pipit and lark song, and there have been buzzards, ravens and ponies to keep me company. Now there may be military shooting going on and Okemont Hill and the route to the mysterious Cranmere Pool blocked.

To the west of the moor lie Tavistock and Brentor – the latter an 1,100-foot hill, crowned by a church visible for miles and described by Tristram Risdon in 1625 as 'full bleak, and weather beaten, all alone, as it were forsaken'. I climbed alone up the volcanic cone, where the yarrow was in bloom and the linnets were singing their inconsequential songs, to the Church of St Michael of the Rock. The dedication to St Michael is common for churches built on high places. From the church porch I had a glorious panoramic view of Dartmoor. Below me was Lydford, now a village but once one of the most important Devonshire towns. The Norman keep of the castle was formerly the seat of the Stannary Court and Prison which dealt with the tinminers, somewhat roughly if this early seventeenth century verse is to be believed:

> I oft have heard of Lydford Law,
> How in the morn they hang and draw,
> And sit in judgement after.

Close by the castle are the wooded gorge through which the River Lyd swirls and tumbles, where herons and dippers can be seen fishing, and the church, dedicated to St Petrock, the most popular of the Celtic saints of western Britain, where the robins and hedge-sparrows sing among the gravestones. In the churchyard by the Watchmaker's Tomb with its strange inscriptions and analogies I listened to a song thrush while collared doves crooned in the trees beyond. On the moor above Lydford there is a wealth of cairns, hut circles, camps and alignments of stones like that near Merrivale which I walked along one day with a chain of skylark song overhead as I compared it with the *alignements* that I had seen at Carnac in Brittany. There are other stone rows like those near Challacombe Down, and the region is richly sprinkled with prehistoric remains.

Not far from Widecombe and Postbridge is Grimspound of the late Bronze Age, with its couple of dozen huts encompassed by a wall and serviced by a stream that runs through the pound. I watched a pied wagtail catching flies near the somewhat elaborate entrance.

43

Set among the stones and hut circles and other traces of early man on the moor is Princetown with its prison, whose sullen greyness sends a shudder down my spine whenever I see it, while the gaunt lattice-work of the BBC's television transmitter mast rises sharply up into the sodden clouds. Granite forms the scenery and builds the churches and some of the most impressive farmhouses in England. Here the moor is covered with grass, heather and bogs. Rivers sparkle and bubble through the peaty grass and over the granite, slipping beneath ancient clapper bridges like that at Postbridge, tumbling over rocks as at Becka Falls and flowing under arches as at Fingle Bridge. These are the haunts of grey wagtails and dippers, and piratical mink hunt for fish and waterfowl along the River Teign. Both the Tavy and the Dart have salmon and sea trout runs but the brown trout living in the more acid water of the small streams are small; experts say that the best all-round fly for Dartmoor streams is without doubt the Pheasant Tail.

Dartmoor also has its woods. The border country near Bovey Tracey is well wooded and includes some of the best of the Dartmoor native broad-leaved trees. Holne Woods on the River Dart are full of song birds. Yarner Wood is a National Nature Reserve. The Forestry Commission too have planted conifers on the moor. Some of their Sitka spruce plantations may hold breeding siskins. These first appeared at Bellever near Postbridge in 1957 and then at Soussons Down and in 1970 I saw siskins near Two Bridges. These new woodlands have also been colonized by redpolls as well as coal tits, goldcrests, chaffinches and willow warblers. Many Dartmoor buzzard nests are built in conifers despite the presence nearby of broad-leaved trees. These moorland predators feed on rabbits, rats, occasional woodmice, voles, shrews, frogs and toads which all live on Dartmoor.

It is, however, the oakwoods of Dartmoor which appeal especially to me. Above Two Bridges and Crockern Tor, where the tinners' parliament used to meet, and on the east bank of a small stream, is Wistman's Wood, standing at more than 1,200 feet above sea level. It is a pedunculate oakwood growing on block scree known locally as clitter. One day as I made my way towards the wood two hounds from a local hunting pack detached themselves, came disarmingly up to me to nuzzle my hand before turning away to rend ferociously the skull of a sheep not long dead. A stonechat gave the alarm and linnets twittered nervously as I advanced towards the wood. I walked over rough areas of matgrass and purple moorgrass with bushes of western gorse flowering on the west bank as I traversed the valley of the West Dart. Wistman's Wood is a fragment of the old native oakwoods that once covered the slopes of the moor. Dwarf pedunculate oaks are rooted between the granite boulders and often lie horizontally or at sharp angles. They grow only to heights of about fifteen feet but their canopies have a wide span. I found that the trees were covered with mosses, liverworts and a few ferns. Among them were several rowans, hollies and sallows, ivy and honeysuckle and there were considerable amounts of bilberry, great woodrush and wood sorrel under the trees, just as in the Killarney oakwoods which I shall describe in the next chapter. In fact, Wistman's Wood is rather like a sessile oakwood in northern Britain. In summer I found tree pipits in song and have recorded meadow pipits, woodpigeons, cuckoos, mistle thrushes and blackbirds, robins, wrens, chaffinches and carrion crows. The trees just manage to exist under the extremely windy conditions in which they live – a relic of the old oakwoods that were destroyed largely by the tin-miners. There are other fragments of oakwood at Black Tor Copse near Okehampton and Piles Wood by Ivybridge. Both Wistman's Wood and Black Tor Copse are Forest Nature Reserves. In autumn the woods may shelter treecreepers, wrens, robins, blue and coal tits, goldcrests and sometimes woodcock.

I have roamed the moor from Okehampton to Ivybridge and Tavistock to Widecombe-in-the-Moor – white under December blizzards, brown and blue on a summer's morning and gold and magenta in the autumn sun. There is so much to see and hear. I have listened to ring ouzels whistling deep in the moorland ravines, to magpies rattling among the yews, cedars and magnolias of Drake's home at Buckland Abbey, to pied wagtails singing in the main street of Chagford and a redstart delivering his song by Gidleigh Castle. I have watched the lively Dartmoor ponies being rounded up, as the migrating ring ouzels passed overhead, and then being sold in October at the Ashburton pony fair where they made a bird-like chorus of whinnying sounds. The River Tamar separates Cornwall from Devon and it flows past the ancient river ports of Calstock and Morewellham – the latter a copper ore port and the former hub of the communications network of the upper Tamar Valley, and now with its empty docks, granite bollards and stores one of the most evocative places in Devon. The Tamar meanders on under rocky cliffs and woods of oaks, birch and sycamore and then between muddy banks where mallard, gulls, common sandpipers and greenshanks gather in October, while cormorants perch in the leafy trees nearby. As I have journeyed down the Tamar, aboard the *Sepia* – the research vessel of the Marine Biological Association in Plymouth – under the 1963 road bridge and Brunel's iron bridge at Saltash, I have seen skylarks and meadow pipits migrating up from the south-east over the frigates and other naval ships.

This part of England is a paradise for the marine biologist, with fine beaches, sea caves and countless rock pools. The red Devonian rocks give way to those that have been transformed by heat and pressure and resist the storms and constant pounding of the sea. I have explored the rock pools at Wembury and Lannacombe, collecting and pressing seaweeds, examining the underwater world of the sponges, corals, beadlet and snakelocks anemones, the whelks and acorn barnacles, the prawns and hermit crabs, the blennies and the butterfish and the rocklings, and filming for television all these delights as well as the microscopic world of plankton.

The coastal scenery is surely more varied here than on any comparable strip of English shoreline. There is the ruined village of Hallsands where erosion brought about by man's unthinking removal of more than 600,000 tons of shingle has ended the life of one human community. At Start Point not far away the herring gulls nest on the ledges and rock shelves above the seas. One morning I stood and watched fulmars, cormorants and shags on the cliffs, rock pipits flitting from outcrop to outcrop, displaying oystercatchers piping and trilling among the rock pools, and razorbills and guillemots on the sea below. On the clifftop three hares were running in circles in their spring ecstasy and a buzzard was quartering the rough ground looking for small prey. Many times the brown bird of prey hovered with beating wings before passing on to a new hunting ground. A raven called behind me and a peregrine with alternately winnowing and gliding wings passed out over the sea – a dark menacing crescent.

To the north of Start Point and Hallsands is Slapton Ley – a large freshwater lake which runs parallel to the sea and is separated from it by a bank of sand and shingle where the D-Day landings were rehearsed. The sands are long and for the beachcomber full of interest, and one of my prized possessions is a bottle encrusted with acorn barnacles that I found among the crab shells and whelk egg-cases on the beach. The pool is about 300 acres in extent with reed-filled bays where coot, moorhens, sedge and reed warblers breed. In early April hundreds of swallows and house martins can be observed above the Ley where, as I once saw, a black-throated diver was beginning to assume his summer plumage. Linnets, stonechats and reed buntings occur in the scrubby cover along the sands and the shore of the Ley. It is a good place to watch birds arriving and I have seen cuckoos making their first landfall among the wheatears. A duck mallard with ten ducklings came in from the sea one day, having probably made a mile of sea crossing from her nest near Start Point. She kept a wary eye on the buzzard circling high above the beach on broad stiff wings.

The Ley is a good place to see insects and especially dragonflies. Among its more unusual plants I have listed small-flowered buttercup, sea radish and common broomrape among the sands, Kaffir fig and Russian comfrey near Torcross, three-veined sandwort from the Wood, and from the Ley wavy bittercress, small-flowered catchfly, parsley piert and wild celery, field and wild madder, curly pondweed, grey sedge and greyish bulrush, *Arenaria balearica*, and two rare plants – strapwort (a small prostrate member of the pink family) and snowflake.

In April 1961 I stood on the bridge near the sluice where the Ley spills over into a tunnel that descends to the beach and the sea. Below me in the water dark clouds of elvers – young eels which having been spawned in the Sargasso Sea had been drifted on the currents and reached Devon – were kicking, jerking and twisting their way up the stream, some being lost in the bubbling eddies and washed back into the tunnel, others reaching the calm shelter of the Ley after an homeric struggle. There is much to see of nature's variety on and around these western moors with their long history and legends, their coombs and valleys, great high uplands and rocky outcrops, bubbling streams and cataracts. Enchanted lands and islands lie further west – Cornwall, also with its granite, and the Isles of Scilly. I have paused on the edge of Dozmary Pool on Bodmin Moor – the reputed home of the Lady of the Lake where Sir Bedivere saw 'the ripple washing in the reeds' before returning Excalibur to its home. I saw herring gulls bathing in the fresh water and calling antiphonally, and here with Brown Willy commanding the far horizon I found that the romantic historian-naturalist could let his imagination wander until legend overcame reality.

But reality for these western moors may be very different. Dartmoor is under threat and this is reflected, for example, in stark white pyramids of china clay, while stretches of moorland on Exmoor are still being brought under the plough despite Lord Porchester's Report and the opposition of the Countryside Commission, which was set up as long ago as 1949 to select the most suitable areas 'for the purpose of preserving and enhancing their natural beauty and promoting their enjoyment by the public' – the key to the National Parks of England and Wales.

3. The Green Isle

My first view of Erin's green isle was from an aircraft swinging round over Strangford Lough on its approach to Nutt's Corner, where it landed in an explosion of fast-sprinting hares and flocks of golden plover. The second approach was a more leisurely one by the sea crossing from Stranraer to Larne – a manufacturing town set in a ring of hills from which one can look back to the tiny cone of Ailsa Craig, whitened by its thousands of gannets. From Antrim I have been able to tour Ireland from Rathlin Island in the north-east to Cape Clear Island in the south-west, and from Dun Laoghaire in the east to Connemara in the far west. The landscapes of Ireland are, in fact, closely linked in their rocks and structure with those of Britain, the major features of the Scottish Highlands, the Uplands and Wales being continued into the Green Isle. The separation of the two islands by the Irish Sea is quite recent. Much of the island's heart is covered by a great expanse of Carboniferous limestone, broken from time to time by outcrops of earlier date, yet a great deal of the countryside is overlain with boulder clays and peat so that the limestone rarely appears at the surface.

The island's centre is a vast plain or plateau, and in conditions of almost constant moisture mosses and bogs are frequent, while many of the expanses of water are probably glacial lakes. There are basaltic lavas in Antrim, granite in the Mountains of Mourne, and Ireland is formed largely from palaeozoic rocks. In Donegal, Kerry, Mayo and Connemara conditions are reminiscent of the Scottish Highlands, and in Down and Armagh of the southern uplands of Scotland, while the landscapes of Scotland continue less obviously into the Irish Plain. In Miocene times enormous sheets of molten lava poured out over Antrim, producing in the process of cooling, basaltic hexagons and columns, and in their vertical jointing a landscape of black cliffs and glens. My tour of Ireland and my investigation of its flora and fauna start along the Antrim coast north of Larne, but before I embark upon them a small digression may be helpful.

Northern Ireland shares with Scotland its geological history and its separation is political rather than geographical. Ireland was once rich in deciduous woodland but from the early seventeenth century onwards land was cleared for agriculture, and by 1700 the old forests of Ireland had been reduced almost to nothing. Some kinds of woodland bird, such as the three woodpeckers, the marsh and willow tits, nuthatch, nightingale, lesser whitethroat, tree pipit and the pied flycatcher are missing as breeding birds in Ireland. There are no snakes, slow-worms, sand lizards, warty and palmate newts or common toads among the reptiles and amphibians. Among the mammals moles are absent as well as common and water shrews, greater horseshoe and noctule bats, barbastelles, short-tailed and water voles, yellow-necked mice, harvest mice, common dormice, weasels and roe deer. In the butterflies notable absentees are the mountain ringlet, marbled white, Scotch argus, heath fritillary, comma, purple emperor, white admiral, silver-studded blue, small skipper and other insects.

To some extent the wildlife of Ireland is an impoverished one brought about, either by the earlier separation from Britain after the Ice Age and before certain north-westward moving species could reach Ireland, or by conditions of terrain and climate. This does not mean that the naturalist cannot enjoy himself in Ulster and Eire, and some of my most interesting wildlife experiences have been in the woods, along the rivers and on the hills and islands of both countries.

My tour of Ireland begins with a drive north from Larne up the coast road of Antrim – one of the most fascinating roads in the whole of western Europe. The scenery itself is unusual since it is composed of layers of basaltic lavas poured out some 60 million years ago over beds of chalk. The high landscape to the west is one of stepped land forms, penetrated by the famous deep Glens of Antrim, each with a splendidly melodious name. Patches of bright green farmland push their way up the glens and, circled by stone walls, have come to be known as 'ladder farms'. On the slopes peat is cut from the many bogs that have formed over the boulder clays left by the ice sheets. As I drove up to Glenarm there was a diversion in the road, for the chalk rests on slippery clay and a giant landslip had closed the coast road. Indeed one settlement between Glenarm and Carnlough is still known as 'the slipping village'. I paused to watch a corncrake that had walked boldly out into the roadway and on my approach just stood in the grass by the verge.

The wooded grounds of Glenarm Castle were full of bird songs. Off Garron Point the herring gulls were soaring, and then the road carried me along the broad sweep of Red Bay, where the red rocks contrasted with the chalk that I had been seeing for the past few miles. High above me rose the Garron Plateau – a world of heather, grey pools and mist where red grouse, curlew and wheatears had their homes. The Vale of Glenariff was enriched with streams and delightful falls and cascades, where I found dippers, grey wagtails and common sandpipers nesting, while the woods were full of crows and singing willow warblers. I could hear more curlews on the slopes above, rising and falling in their displays, and a stonechat was giving his metallic alarm on the lower rocks of Tievebulliagh, whose dark blue-grey 'porcellanite' was made into stone axes for export to Scotland and England. Cushendall – 'the darlint village' just as it was described in *Juno and the Paycock* – I found delightful, and there was, in close proximity, a naturalist's paradise of woodlands, glens, waterfalls and ocean.

At Cushendun I took the coast road along the clifftop with delightful views across to Jura and later Rathlin Island, but I also made a point of seeing the lonely lough of Loughareema, called by Moira O'Neill, famous for her *Songs of the Glens of Antrim*, 'the Fairy Lough', and here a single golden plover was standing quietly by the water's edge. Ahead of me was Fair Head or Benmore – a towering 600-foot promontory among whose mighty rocks wild goats climb and fulmars cackle and on whose basalt cliffs a pair of golden eagles nested between 1953 and 1960. For me there were no eagles – just a buzzard exploring the rising currents of air.

I called in at Ballycastle with its sycamore trees and flowery gardens and then at Carrick-a-Rede walked over the frail swaying rope bridge, put up each spring and dismantled each autumn, which links the mainland with a kidney-shaped island of turf-covered basalt. Here from the rope bridge across a deep gorge through which the salmon pass I looked down on gulls and kittiwakes stacked on the ledges, while shags and seals were swimming on the sea below. I journeyed to the Giant's Causeway, making my half-mile walk to the accompaniment of wren and rock pipit songs. Here were grey forbidding humps of columnar basalt, not as impressive as Staffa, and covered in places with a thin green carpet of grass. Where the Grand Causeway slopes down to the sea the exposed tops of the hexagonal columns of basalt form a kind of pavement or contoured pyramids like a church organ. Just around the corner is Port Ballintrae, where

I stayed several times to enjoy the setting and the sea; nearby lie the Spanish galleon *Gerona* and the gaunt stone gables, round towers and grey walls of Dunluce Castle – the former seat of the Earls of Antrim – dating back to the thirteenth century and overhanging the sea. This north coast of Antrim looks out to Rathlin Island. From Bally-castle it lies as a long, dark enticing line on the horizon – a mighty jagged reef and for me one of the most delightful spots on earth.

I sailed out of Ballycastle in the *St Frances* across the Brochan – 'the boiling porridge pot' – that rough and sometimes impassable piece of ocean that separates Rathlin by some seven miles from the main-land. The high-prowed open boat was making for the little harbour in Church Bay, which nestles in the trigger-guard of this pistol-shaped island pointing westwards away from the Mull of Kintyre and into the Atlantic. As I approached the island I could see that white buttresses of limestone, split into cracks and caves, supported the steep black basalt cliffs on top of the island. Charles Kingsley, impressed by Rathlin's cliffs of black basalt and snowy limestone, thought that it looked 'like a drowned magpie'. Church Bay is ringed with high cliffs that look down on the small solid church with its square tower and pointed windows, the red-roofed cottages and the long stone manorhouse above its grey wall. The harbour has two piers, silver sands where common and black-headed gulls were stamping to bring up the worms and choughs searched through the black tufts of seaweed, a ramp where several small boats were hauled up, a hall, school and a post office. On my first visit I was taken around the island in an unsprung, back-breaking trailer drawn by a tractor but I returned later to spend whole days travelling and walking around the island, which I came to love. Some of its wildlife treasures have now found safety, I am glad to say, with the Royal Society for the Protection of Birds.

There is a stony road on Rathlin which leads to Mr and Mrs McCurdy's house where I stayed. Its windows, from which I looked out over Church Bay, were surrounded with green-painted bricks. A veronica hedge gave shelter at the front and the dog from next door, Tuskar, took up his guard post on one of the tall concrete gate-posts. Around the house were green fields white with daisies where black cattle dropped their calves and sheep grazed, confined by the walls built of lumps of black basalt or white limestone rocks. The road ran on to the south past little houses and lochans, known as loughs, which were bright with bog bean and water lilies, where mallard and tufted duck, coots and moorhens were swimming. Common gulls wailed and snipe chippered away near the water. The side of this track shone bright gold in the sun with gorse, broom and tormentil, and I spotted lady's mantle, burnet roses, red lousewort and orchids. Corncrakes craked away invisibly in the grass, their voices sometimes loud, sometimes ventriloquially soft. It was warm on the little farms, heathery uplands and along the dusty, flinty road.

As I came closer to Rue Point and its automatic lighthouse, I could see sheltered bays of clear water and a fringe of seaweed where shelduck and their flotillas of piping young were dibbling, eiders swam and dived, and oystercatchers trilled and fluted in display. Far overhead a buzzard mewed and the gruff calls of the choughs came floating on the summer breeze. I used to take up my position by some deserted huts near the lighthouse. To the south I could see the rounded slopes of Knocklayd above the white sandy beach at Ballycastle and the white gash in the cliffs above where men were quarrying stone. I clambered up a steep slope to the west of Rue Point among countless spikes of spotted orchid, clusters of birdsfoot trefoil, clumps of flowering sea pink and the blue stars of vernal squills. Wheatears were calling and, as I peered over the cliffs, I looked down a precipice over the sea on which rafts of guillemots and razorbills were floating and from which puffins growled at me in mild alarm. Herring gulls called anxiously at my appearance above them and a rock pipit was singing from a tiny ledge. Some racing pigeons also flapped up in a tired desultory fashion but these were all to fall victim to the peregrines in the next couple of days.

Near the church a track climbs and swings west and here I picked up a series of worked grey-white prehistoric scrapers – the first signs of man's association with the island which was known to Pliny and Ptolemy, where Robert the Bruce stayed and watched the spider and whose cave I visited, and where Marconi flashed his faint Morse signals to Ballycastle at the end of the last century. Here on the track there were more ebony, red-billed choughs – one of the island's most exciting bird species. Bushes of fuchsia and veronica appeared, and a salt-blasted clump of stunted conifers – an experiment by the Forestry Division of the Ministry of Agriculture.

The coastal scenery of Rathlin is magnificent, with columns of basalt rising steeply up from the sea. I could stand on two columns three feet apart and look down between my toes more than 400 feet to the narrow strip of beach below. On the limestone cliffs kittiwakes, choughs and black guillemots were nesting and a buzzard used a white buttress of rock as a plucking station. At the West Lighthouse at the end of the track I saw a huge steep slope of concrete that fell away in breath-stopping leaps and bounds to the lantern on its man-made ledge below me. Great sea-girt stacks rose up like ziggurats or step pyramids and here there were seabirds in their thousands. On the lower ledges herring gulls and great black-backed gulls were incubating or tending young and rock pipits were launching themselves upwards into their song flights. Higher up on tiny ledges and rocky perches above the breaking sea were cities of kittiwakes. Above them again and before the basalt rose sheer once more were green turf-covered slopes where heavy-billed puffins sat in or by their nesting burrows. A few fulmars were breeding on the slopes and I could see black-and-white razorbills in nearby crevices and cracks. The top of the nearest stack was so thick with guillemots that it looked like a giant iced cake dotted with tiny bits of broken chocolate. The guillemots cawed and moaned – some of Rathlin's most evocative sounds – while I pondered the past history of the island – its potato famine and its decreasing population. It was Lettice Stevenson who wrote of 'Rathlin, Rathlin, gem of the Northern Sea' and who summed up its charm and interest for the naturalist.

> There's Ushet Lake and Claggan Lake and heather on the hills,
> And fearsome cliffs above the sea where sit the razorbills,
> And puffins, guillemots and gulls all scream as they fly by,
> The 'East Light' and the 'West Light' watch the great ships
>   moving by.

That is just one part, albeit a delectable part, of Northern Ireland. There are also strips of woodland set in a lush green landscape with its many hedgerow trees and flowering hawthorns. At Breen there is one piece of sessile oakwood with some birch and rowan where rushes, bracken, bramble and wood sorrel adorn the Antrim countryside. Here mistle thrushes were twice as common as blackbirds, for there was little in the way of a shrub layer. I watched buzzards circling above the Ulster landscape, and long-eared owls hunted through the wood in daylight while chiffchaffs and cuckoos were common. In Northern Ireland Co. Fermanagh has the most woodland, followed by Londonderry, Tyrone, Antrim, Down and Armagh. The State planting of trees has grown in recent years and of this about 90 per cent consists of conifers, but broad-leaved trees are generally planted in mixture with the conifers so that the latter can be thinned out to allow hardwoods to mature. At Port Glenone west of Ballymena I visited one plantation of ash, birch, rowan and sycamore mixed with pine, larch and spruce where foxes had an earth and there were at least eighteen different species of bird including many redpolls. In Co. Down many conifers have been set on the lower slopes of the Mourne Mountains. In the younger forest plantations – that is, those under ten years of age – meadow pipits can be heard singing in the company of willow and grasshopper warblers, showing the transition in land use from that of heathland to woodland scrub. East of Belfast in the six-to-eight-foot high conifers at Cairnwood sedge warblers were the dominant species.

Mount Stewart on the Ards Peninsula is not far away, and its gardens full of ilex, magnolias and firs with tall cedars, Wellingtonias, beeches, limes and eucalyptus trees resound with birdsong, and the treecreepers excavate roosting holes in the soft bark of the Wellingtonias. The gardens stand on the shores of Strangford Lough where many black-headed gulls now breed, threatening the terns that nest in the same area – common, Arctic, Sandwich and roseate. (Off the mouth of Belfast Lough the Copeland Islands are a favourite haunt of the roseate tern, now showing signs of a catastrophic decline in the British Isles.) In winter Strangford Lough is a marvellous place for birds. One January day near Comber I saw about ten thousand wigeon and thousands of knot as well as smaller numbers of oystercatchers, godwits, redshank and lapwings. Opposite Horse Island there were two parties of Brent geese – the pale-bellied form – about seven hundred altogether, with flocks of curlew and golden plover.

South again of Strangford Lough is Tollymore Forest Park – the first forest of the Government of Northern Ireland to be opened to the public and covering almost 1,200 acres with views of the Mourne Mountains and the Irish Sea at Newcastle. The Park is underlain by rocks of Silurian age – slates and shales – with veins of igneous rock, while boulder drift covers the top. In the mixed broad-leaved forests where I walked I found blackbirds – the commonest birds – and chaffinches, robins, wrens, mistle thrushes, four kinds of tit, goldcrests, hedgesparrows, woodpigeons, chiffchaffs, bullfinches and jays. There are also dippers and sometimes kingfishers along the streams. Foxes and badgers are common, otters occasionally visit the River Shimna and there are pygmy shrews, red squirrels and pine martens. I was walking through a plantation in this region in May 1967 carrying out a bird census when a pine marten emerged from the woodland onto the path and began to lope towards me. I stood perfectly still and this rare and attractive mammal came right up to me, sniffed my boots, looked up and then quietly went on his way! Surely with his rich chestnut-brown colour and creamy throat patch or bib the pine marten is one of our most attractive mammals.

Farther south still, on Carlingford Lough 'where the Mountains of Mourne run down to the sea' lies another favourite haunt of mine – Rostrevor. It has been described as an English village in a Norwegian fjord and it is set against a backcloth of wooded hills. The Fairy Glen is a luxuriant wooded valley through which the Kilbroney River flows, and here I have watched dippers, grey wagtails and Irish jays. From Cloghmore Glen mixed stands of conifers climb up the hills like a piece of carpet, wearing thin near the summits, and this is great pony trekking country. On the high ground a few pairs of ring ouzels still hang on. Near Newry I walked along a strip of mixed broad-leaved wood outside the coniferous Narrow Water Forest; this was a mass of bluebells and was full of singing chiffchaffs, willow warblers, thrushes, chaffinches, goldfinches, robins and wrens.

One of the most interesting natural features of Northern Ireland are the reed-fringed loughs, rich in water birds. Lough Neagh consists really of shallow water collected in the saucer formed by a collapsed lava plateau. Large numbers of great crested grebes nest in the reedbeds, with coot, moorhen and occasional water rails. The water contains trout, eels and Lough Neagh pollan, the latter spawning on hard firm ground on the bottom of the lake. In winter Lough Neagh is visited by thousands of tufted duck, pochard and goldeneye, which gather on its 150 square miles of freshwater lake. Close canopy birchwoods grow around the Lough and shelter rowans, sallows, ivy, brambles, ferns and mosses while the birds I have seen when I have sauntered through these woodlands include mistle and song thrushes, blackbirds, wrens, willow warblers, chiffchaffs, chaffinches, robins and many other kinds of song bird, while in the birch and sallow scrub that I investigated there were also grasshopper warblers, greenfinches and goldfinches. Once the region provided oak for British battleships but by 1700 the area of forest had almost disappeared. Not far from Lough Neagh is an area of pits and excavations, some full of water, others of sedgy vegetation where black-headed gulls and common terns breed noisily and conspicuously. Driving around in this part of the world can be hazardous, since uncontrolled black-and-white sheepdogs lie in wait for cars in the roadside grass and rush out to snap at and even bite the tyres of the passing vehicles!

West of Lough Neagh Northern Ireland presents a range of different landscapes from the grandeur of the Sperrin Mountains to the noble woods of Baronscourt, seat of the Dukes of Abercorn, with its fine oaks, beeches, alders, ashes and rhododendrons, in which I found masses of bluebells and strong-smelling ramsons and many woodland birds, and to the lakeland countryside of Lough Erne. I have made my bases in Omagh, which was full of redpolls, and Dromore, and from here I was able to explore the Gortin Glen Forest Park, Seskinore and Upper and Lower Lough Erne. Gortin Glen is typical of the West Sperrins and is an overflow valley of one of the great glacial lakes; its base is one of schist, gneiss and quartzite with blanket bog over much of its area and such typical plants as bilberry, heather and bog myrtle with some birch and rowan. The trees that have been planted are mainly Sitka spruce with lodgepole pine, larch, spruce and silver fir. One of the Park's great attractions is the scenic drive, affording fine views of the forest and the Strule Valley with its trout and pearl mussels. Foxes, stoats, badgers – I met a tame one at the Forester's Lodge – introduced red, not grey, squirrels, hares and rabbits are very common and there are a few Sika deer, while the commonest gamebird is the red grouse. There were more Irish grouse near the road from Fintona to Five Mile Town. Meadow pipits, skylarks and redpolls were especially common at Gortin and, like most of the woods I saw, these held sparrowhawks too. I was very pleased to see so many of these birds of prey in Ulster. From the Park one looks out across gentle undulations of grey-green hills, white houses lying under cloud shadow and pools of bright sunshine – Northern Ireland at its most typical and best.

The chief town of Co. Fermanagh – Enniskillen – lies on an island between channels of the River Erne which divides Upper from Lower Lough Erne. Upper Lough Erne is a lake of wooded islands, round towers and ivy-clad castles. In one secluded bay I have watched and tape-recorded parties of common scoters – commonly sea ducks – with several of the jet black drakes scuttling across the water, ploughing a watery furrow with their bills, raising themselves above the surface with heads and tails up, giving vent to a most musical call. This quiet inlet with its enveloping trees resounded with melodious pipings and rapid rattles as the drakes in their display clattered with black wings across the Lough. The horizon was charmingly wooded with enticing islets covered in trees, set among patchworked hills. On the southern shore of Lower Lough Erne I walked among quite tall ash trees with a rich layer of flowers underneath. Around the two lakes I could hear blackcaps and garden warblers in song and the contralto phrases of the latter were audible too from the wooded islets. In Ireland garden warblers seem to need a lake in their habitat and are largely confined to the Shannon Lakes, Lower Lough Erne, Lough Neagh and the Cavan lakes. At Castle Caldwell on Lower Lough Erne I found spruce plantations growing with broad-leaved trees, especially oak, beech and ash, and both here and at Castle Archdale on the opposite bank chiffchaffs and willow warblers were the co-dominant species.

My first introduction to Eire was by way of the rugged fjord-indented coast of Donegal with its thatched white cottages and stone field walls. After entering the Republic at Strabane I motored through soft country to Letterkenny and Milford with its blue lough, dotted with tiny wooded islets, set in low purple hills like dumplings, with rough wastelands illuminated by brilliant patches of yellow gorse. Milford is an angler's resort and I could guess why. Around Mulroy Bay were strips of sessile oak scrub with ash, alder, hazel and some birch, holly, rowan and gorse. Below Muckish Mountain I turned off to wander through another wood of spruce, sycamore and ash not far from Dunfanaghy. In another region of birch scrub and alder there were many willow warblers and chiffchaffs in song – birds I came very much to associate with Irish woodlands – as well as robins, wrens and other songbirds. Close by is Horn Head where one can see breeding gulls and auks. In this fascinating and wild north-western corner of Ireland Errigal Mountain, all 2,466 feet of it, rises pyramidal and impressive from a scene of rock folds, stone walls and whitewashed farming settlements. Not far away is Bloody Foreland – once a schoolboy joke in an Atlas – but in reality a long thin promontory like a dead human finger, crossed by straight and zigzag parallel stone walls, grey wastes and sturdy white farms, some of whose outbuildings with their patches of missing tiles looked like lace. The scenery was dramatic and arresting. I paused at Maas near Gweebarra Bay to walk in a wood of sycamores and ash trees which held a large rookery and many chaffinches, blackbirds, song thrushes, robins, wrens and a couple of chiffchaffs. But the bird populations were becoming increasingly predictable. Donegal has a landscape perhaps unsurpassed anywhere in Ireland with its rugged cliffs, long fjord-like inlets, broad bays, deep glens, heaths and moors, bogs, lakes and mountains. The general barrenness of the hills of Donegal arises from acid metamorphic rocks which form its mountains, while quartzites, mica-schists and granite come to the surface and from the quartzite have weathered many of the hills which rise up above the schist and granite. The soils of Donegal consist largely of rock detritus and some patches of glacial drift and, although lowland bog is not common, mountain bog is frequent. It reminds me in parts of the Scottish Highlands. The birds range from the ravens, merlins, wheatears, twites and red grouse of the moors to the snipe, common sandpipers of its lakes, the stonechats and choughs of the coast and the woodland and scrub birds to which I have already referred. There is a tiny population of red-throated divers which has clung to its north-west Ireland toehold since 1884, and there are storm petrels on Roaninish and Rathlin O'Birne Islands.

I brought my tour to an end at the market town of Donegal on the River Eske. Here were rows of slated white houses, a quay, a church with a spire like an English parish church, a Jacobean ruin of a castle, a wide sweep of river fringed with broad-leaved trees that frame an arched bridge and a centre placed in a triangle called 'the Diamond'. This was once the capital of the land of the O'Donnells – the lords of Terch'onaill.

My route then took me south through Ballyshannon and the resort of Bundoran and then on below a plateau of grey indented hills ending in the bluff of Ben Bulben, which sits above like the prow of a ship or a cob mute swan's beak and knob. Below it lies the village of Drumcliff where W. B. Yeats is buried; a spotted flycatcher was sitting above the poet's own words on his grave:

> Cast a cold Eye
> On Life, on Death,
> Horseman, pass by!

I visited Sligo – 'the shelly river' – with its Abbey and delightful arched and pillared cloisters, and turned east to Lough Gill to see for the first time Yeats's Lake Isle of Innisfree 'full of the linnet's wings'. This elegant wooded islet rising from deep blue waters was ringed with the double calls of cuckoos and the clear songs of blackbirds, song thrushes, and many willow warblers and chiffchaffs – those inseparable companions of the Irish woodlands. Round about are delightful parklands and municipally owned estates with tall broad-leaved trees and evergreens. Yew and rock whitebeam grow here and I saw several strawberry trees in the limestone rock. The superb Ben Bulben hills are formed of Carboniferous limestone; they rise to over two thousand feet, from acid farmland and bog to peat-covered limestone on the summits, and on the rock outcrops grow mountain bedstraw, milkwort, mossy saxifrage and the rare Irish sandwort with its bright white flowers.

Then I backtracked on my route to visit Ballina and the mixed oakwoods around Lough Conn; here the countryside embraces the Ox Mountains, formed from granite and overlain with gneiss and schist and the quartzite Nephin Beg Range west of the Lough. Lough Conn is noted for its trout and so-called gillaroos – a variety with many spots. At Pontoon a bridge spans the short stream between Lough Conn and Lough Cullin and here I lunched under a canopy of mature sessile oaks growing amongst spruce trees and with a shrub layer underneath of fine beech and birch; my alfresco meal on a warm still day was enhanced by a chorus of the inevitable warblers, thrushes, coal tits, goldcrests, chaffinches, robins, wrens and the coos of woodpigeons. I watched a fox cross a field and scale a wall with a single bound before disappearing into a hole under the roadway. Every few miles in Co. Mayo I came across a priest deep in meditation and soon learned not to inquire the way!

From Mayo my route ran through Boyle, Carrick-on-Shannon and over the great Ice Age deposits of the Central Plain. Beeches and ash trees often graced and even dominated the hawthorn hedges, where yellowhammers churned out their simple repetitive ditties. Near Roosky in Co. Offaly I took a census of the birds in a close-canopy birchwood to find that wrens were the commonest birds, followed by robins. It was a great delight to walk among the graceful birch trees with the sun warm on my back and the birds singing all round me and butterflies welcoming the sunlight. In Ireland woods of ash are not very common but at Kilbeggan on a glacial moraine known as the Big Esker I was able to explore quite a large one. There are others on the hills and plateaux of Sligo, Galway and Clare but ash was probably never very common in Ireland. In the Kilbeggan wood there was some birch and beech with hazel and thorn, but the floor was beautiful and rich with bluebells, bugle, sanicle, arum, lady's mantle and two orchids – common spotted and twayblade. Blackbirds and willow warblers were the dominant singers.

It is not much more than a stone's throw from Kilbeggan to Tullamore, which has a rare pedunculate oakwood, whereas most Irish oakwoods are of the sessile kind. This cross-country journey through rich, green farmlands, past hedges and grassland was to bring me to the high barren moorlands that stretch south from Dublin to Co. Wicklow – a granite ridge with schists, slates and shales in the richly wooded valleys which hold such gems as Powers-court, with its ornamental gardens and Deer Park, waterfall and avenue of beech trees, and the Glens of the Downs, Dunran, the Devil's Glen with its stream and rocky sides, Clara where I heard redstarts in song, Avoca and Shillelagh.

Glendalough – the Valley of the Two Lakes – is an enchanted spot and only a day's journey from Dublin. In a deep solitary wooded glen, approached across the Wicklow Mountains – treeless, bare, covered in deer sedge and purple in the sun – lie the remains of several churches which have existed for more than twelve hundred years and a tall round tower that has stood there for nearly a millennium. Here the rain-washed air gives the glen an ethereally soft light, so appropriate to some of the oldest Christian buildings in Europe and associated with the sixth century St Kevin. I walked round St Kevin's Kitchen – a church with a steeply pitched roof of stone and a round-tower of a belfry, nestling among isolated conifers and gravestones crowned with Celtic crosses. The valley is full of trees and the glensides come close together so that there is an indescribable air of enchantment and mystery. Described by Sir Walter Scott as 'the inestimably singular scene of Irish antiquity', Glendalough was to Thackeray 'a delightful place' with its 'little lake, and little fords across it, surrounded by little islands, where there are all sorts of fantastic little old chapels, and graveyards: or again, into little breaks and shrubberies, where small rivers are crossing over little rocks, splashing and jumping, and singing as loud as ever they can'. Here I caught glimpses of grey wagtails and a dipper, both of whose watery requirements are rather similar. In the sessile oakwoods that clothed the valley sides there were blackcaps in glorious song, Irish jays and a redstart – a rare bird in Ireland but this rarity is not due to a lack of mature oak trees. I found another pair of redstarts in a Wicklow beechwood where a wood warbler was singing – another great rarity in Ireland.

I was sad to leave Glendalough but I still had to visit the west and south-west of Eire, and this was to be achieved by a circuitous route. To the south Co. Wexford was quite well wooded with broad river valleys, scrub and conifer plantations. Again chiffchaffs and willow warblers were singing in every copse and thick hedgerow with stonechats and meadow pipits in the younger plantations and gold-crests and coal tits in the older, just as I had seen them in mainland Britain. Wexford Harbour is embraced by the North and South Slobs; the fields and pools are noisy in summer with sedge warblers, larks and pipits but in autumn these areas and the harbour hold large numbers of black-tailed godwits, ruffs, spotted redshanks and other waders, and in winter dabbling ducks and wild geese including Greenland white-fronts and barnacles. There were many tree sparrows along the coast of Wexford, and Great Saltee Island to the south is the summer home of breeding auks, fulmars, kittiwakes, Manx shearwaters, and a couple of hundred pairs of gannets. It is a remarkable place for birds, and its observatory has reported in the past the passage of rare shearwaters, sparrowhawks, kestrels, and the occurrences of rarities like the Alpine swift, lesser short-toed lark, red-rumped swallow, rufous warbler and little bunting.

To the west lay Waterford – a pleasant place to explore – but I travelled on to Carrick-on-Suir, where I watched the swallows over the trout and salmon river. Beyond were the Comeragh Mountains and the mass of Slievenaman and dense conifer woods where the rides were ablaze with bluebells and fringed with beech, rowan, birch and holly; goldcrests and coal tits were abundant but there were also numbers of blackbirds, chaffinches, robins and wrens and there were crossbills too, 'jipping' as they flew away among the pines whose care was the responsibility of Roinn Tailte – the Department of Lands in Eire. I also remember Waterford for the marvellous sight of five hen harriers quartering the same gorse-covered hillside, lending support to the view that this handsome bird of prey has a preference for regions overlying the Old Red Sandstone.

To me south-west Ireland is one of the most attractive and appealing parts of the British Isles. Long arms of greyish purple rock run into the sea to meet and divide the rollers that were born three thousand miles away in the Atlantic; grey stone walls cut off squares of bright green land; gorse blossoms in unbelievable profusion; hedges of fuchsia are bright with their pendent drops of blood and cuckoos can be found along every roadside. From the summits of the sandstone or limestone hills one looks down on huge sea inlets, low grey humps of land and countless islets, while the coast ranges from the white sands of Barley Cove where the sea holly grows to great cliffs, pinnacles, blocks and stacks often standing in a cerulean sea and in summer aglow with sea pinks, splashed by the white explosions of sea that strike the rocky ledges like naval depth charges. In Bantry Bay, the scene of two attempted invasions by the French and a successful one by the oil industry, is Glengarriff – 'the rugged glen' – from which one can take a boat out to Garinish Island or Ilnacullin. There is a martello tower of Napoleonic vintage and an Italian garden where I have studied the reflections in the oblong lily pool of the stone portico and double flight of steps, which are set amongst the greens and reds of cypresses, azaleas and rhododendrons; when I was there the warm soft air was alive with the sibilances of goldcrests and siskins. In the south-west two long bony fingers of land reach out into the Atlantic Ocean, each ending in a great rocky bluff – Sheep's Head and Mizen Head.

At Mizen – the most south-westerly point of Ireland – there is a domed promontory with steep sides where the short turf gives way to bare vertical rock. The cliffs are banded grey and black – the home of choughs, auks, gulls and rock pipits – and from their edge I have looked out across blue-green seas flecked with white waves where basking sharks, small whales, shearwaters and gannets pass by and from which I once saw a huge black-browed albatross sailing past. Mizen is a marvellous place from which to watch the choughs tumbling and calling; once those crows were common along many of our coasts – remember 'the crows and choughs that wing the mid-way air' above the Dover cliffs of King Lear? – but now they have retreated to the west of mainland Britain and Ireland. There are many delightful beaches like that at Rosscarbery, where I found drifts of Irish spurge growing, and one small bay, approached through a wood full of ramsons, which my family and I shared with just one tethered pony, but the water was breathtakingly cold! The roads in this part of the world are full of hobbled horses, cattle and sheep, and Irishmen can be seen grazing their cattle along the grass verges or taking their churn of milk to the local creamery. Here there is time to enjoy the countryside and to talk to people who ask questions and prove that they are genuinely interested in the answers.

This is very much a land of slow-weathering sandstone or limestone and shale. Above Schull rises the 1339-foot Mount Gabriel – a ridge of limestone like a gentler Tryfan pushing up from a plain of rough grazing and rushy ground where the royal fern grows, and a network of small walled fields. As I climbed up the slopes I found lesser bird's foot trefoil and giant butterwort in flower, and curlew and oystercatchers called in the distance. I paused to examine some of the shallow tunnels and drifts that ran into the hillside half way up the ridge and I found half a dozen ancient stone mauls that had once been used to hammer out copper from the veins that reached the surface; nearby Ballydehob was once a copper-mining centre. This is also the land where St Patrick's cabbage grows and near Kenmare the related kidney saxifrage, and there are early purple and marsh orchids, three-cornered leeks and rustyback ferns.

There are many exciting journeys to be made – around the Ring of Kerry, past the mountain ranges of the Slieve Miskish and the Caha and Carrantuohill – Ireland's highest hill – a small peak dominating half a dozen lesser crests and lumps. There are the towering cliffs of Slea Head overlooking the Blasket Islands – seven large islands and many less important rocks lying to the west of the Dingle Peninsula. Steep cliffs are surmounted by grassy slopes with sea campion and sea pink and bog and mountain pasture. Storm petrels breed in the stone walls and Manx shearwaters nest in burrows on the island group, while rock doves still make the Blaskets their home. Above Slea Head rises Mount Brandon, from which St Brendan is said to have sailed to seek 'the land of the blest'. At Gallarus is the Oratory – a stone building like an upturned boat standing with assurance amongst the grass and reached across several stone walls – an early Christian construction of unmortared stones. In the south-west Old Red Sandstone ridges enclosing softer rocks have survived and where the eroded rocks have gone one finds the fjord-like inlets – Dingle, Kenmare, Bantry and Dunmanus. Having been to Rathlin Island at the north-east tip of Ireland I could not resist the call of Cape Clear Island in the extreme south-west off Baltimore Bay.

From Schull the island shows as a long ridge slightly higher in the east than the west. The crossing can be an adventurous one before you enter the tiny harbour with its stone jetty on one side, a grey muddy beach ahead and a steep green bank on the left dropping down to the water's edge. There are long grey cliffs, small stone-walled fields and dwellings, stunted bushes streaming away from the prevailing winds, and the roadsides in summer are decorated with the yellow-bronze spikes of pennywort. I saw fulmars, shags, many gulls, auks, oystercatchers, snipe, rock and stock doves, a few corn-crakes, ravens and hooded crows, choughs, rock pipits and stone-chats. The Island's observatory has listed such rarities as little crakes, ortolan and Lapland buntings, lesser grey shrike, Arctic and Bonelli's warblers, while passages of Balearic and sooty shearwaters and of great, Arctic and pomarine skuas can regularly be seen.

The Emerald Isle is studded with countless lakes of clear water – lakes of the limestone area, lakes associated with the glacial grassy drumlins and the spasmodic temporary ones known as turloughs. The three Lakes of Killarney lie between the Carboniferous lime-stone to the east and north and the Old Red Sandstone to the south and west. At the very attractive Meeting of the Waters Killarney's Upper Lake runs into the Lower and Middle Lakes. Much has been written about Killarney as a beauty spot but less about the attrac-tions for the naturalist. Lake, heath and mountain combine to form an exquisite blend of landscapes. There is a wealth of trees and flowering plants to delight the observant visitor. Around Lough Leane, Ross Island and by Muckross Lake grows the strawberry tree with its white bells of flowers and rough scarlet fruits; it roots itself in the limestone, where it appears on the surface – a truly Lusitanian element in our native flora. On the Lower Devonian sandstones and shales and in the mild, wet conditions sessile oaks are dominant, often growing among partly buried, bryophyte-covered boulders with a secondary growth of holly and yew and an occasional straw-berry tree. Unfortunately rhododendron has spread – 'rampaged' would be a more exact description – and its dense mangrove-like growth has shaded out the ground flora; it certainly interfered with many of my bird counts.

The ground flora is made up largely of bilberry, woodrush and wood spurge. In Derrycunnihy Wood chaffinches, robins, blue tits, goldcrests, wrens and coal tits are the dominant birds. One of the interesting mammals around Killarney is the Sika deer, which was first introduced in 1860, and it was clear that these animals were to some extent preventing the natural regeneration of the oakwoods. On the old Muckross Demesne – the Bourn Vincent National Park – there is a very interesting yew-wood where the trees grow in lime-stone pavement made up of blocks lying obliquely on their sides, with some hazel, oak, ash and other trees; it is a gloomy place with the yews growing in close canopy and casting their shade over some sparse bracken and quite a number of ferns and bryophytes; again, goldcrests, robins, coal tits and chaffinches were common. The islands in Lough Leane are also interesting since tufted ducks and red-breasted mergansers sometimes breed, but the rats are rather a problem.

The most interesting and remarkable limestone flora in Ireland is to be found in that region of Co. Clare known as the Burren. I well remember my first visit to that strange bare range of limestone hills overlooking Galway Bay. I drove through Ennis north-westwards, running into Ennistymon on market day; it took me half an hour to negotiate the ponies, donkeys and cattle that filled the main street, and the Irishmen who insisted on sitting on three occasions on the bonnet of my car to discuss local affairs! Finally I was able to reach Lisdoonvarna and Ballyvaughan. The exposed grey limestone rises here in flat terraces or low hills. Calcicolous grasses, bracken, hawthorn, spindletree and ivy grow in the cracks in the limestone and it is the lower slopes that are richest in plants. I found the rare maidenhair fern, tormentil, spring gentian, shrubby cinquefoil, mountain avens, dense-flowered orchid, limestone bugle as well as bloody cranesbill, yellow-wort, madder, rock rose, stone bramble and other plants that recalled for me other limestone areas – the Peak, Craven and English chalk downlands.

The Burren is a bare place indeed. Oliver Cromwell complained about its lack of water and 'no trees to hang a man'. There is, however, a woody growth that I found quite fascinating – hazel which grows to heights of from two to three feet where the effects of grazing and wind are at their most severe. Under the hazel scrub I saw primroses, wood anemones, celandine and early purple orchid; the commonest birds were wrens followed by whitethroats, robins, chaffinches, skylarks and great tits in that order. Between Ennis and Kilfenora the hazel reaches twenty feet in height and I was able to enter this veritable jungle only by crouching down and following paths that the cattle had smashed through the growth. It was a real delight to explore these hazel brakes, set in daisy-covered fields below the grey parallel ridges of limestone that reached up to the sky, which were full of woodland bird song.

Out at sea in Galway Bay and opposite the Burren are the Aran Islands; these are among the last outposts of land before the Atlantic Ocean stretches away to North America. There are three islands – Inishmore (the Big Island), Inishmaan (the Middle Island) and Inisheer (the South Island). I sailed from Rassaveel across Galway

Bay in the *Queen of Aran*; the sea was a dark grey-blue and behind me lay the mountains of Connemara and the distant smudge on the horizon that was Galway. I was sailing to Kilronan on Inishmore, with its grey harbour and whitewashed houses set about a grassy space where the jaunting cars used to wait for visitors coming off the *Naomh Enna* from Galway, and a green caravan marked TEAS was parked. It was appropriate that at the time of my arrival I had just seen on BBC television Robert Flaherty's great classic film *Man of Aran*. I stayed in a neat white, two-storeyed house with a thatched barn enclosed by a long stone wall. I looked out across more stone houses, thatched outhouses and a sycamore wood to Straw Island, with its lighthouse, and beyond that the grey humps of the Burren. Inishmore itself is about eight miles long and two wide.

Much of the island is bare rock – greyish limestone set in flat pavements full of fossils, rippled slabs like a tideline on the shore, ridges and terraces and loose pieces of ankle-twisting shattered rubble, all riven with cracks, fissures and deep holes. Where great table-shaped rocks lie seemingly firm and rooted, an incautious step will set them sliding or rocking about their pivots. In the crevices I found such ferns as spleenwort, rustyback, hart's tongue and rarer ones such as the delicate maidenhair. Bramble, ivy and honeysuckle sprawl over the rocks, and flowers of saxifrage, wild strawberry, burnet rose, rock rose, eyebright, spring gentian and bloody cranes-bill dance in the sea-wind. I watched sheep and continuously braying donkeys wandering across this stony desert; few birds were singing.

On the south coast of Inishmore rise vertical cliffs on whose rim is balanced the great horseshoe fort of Dun Aengus; three massive stepped stone walls, protected by a *chevaux de frise*, whose sharp-pointed columns of rock set on end like a tank trap were to deter cavalry, looked out across the sea to the 600-foot-high rock wall of the Cliffs of Moher. I watched parties of gannets fishing offshore below Dun Aengus and far below me rose the high whistling of the black guillemots. A cliff fisherman was dangling a line to the sea hundreds of feet below. On this side of Inishmore and far back from the clifftop are black awesome holes and shafts that plunge downwards to the level of the sea below. In one of these was a chough's nest and I could hear the young calling from the darkness. One still evening I walked across the island, reinforced by a glass of velvety draught Guinness, to see the early Christian ruin of Teampall Kieran. On one of the gables a pair of choughs was sitting and calling. I paused to look at the currachs lying like whales cast up on the beach. On the seventh-century monastic cell of Clochan na

65

Carraige – a dry-stone beehive with grass growing on the top and a low square entrance into a stone vault – a stonechat was perched, and wheatears were flitting about Teampall Bhreacain and Teglach Enda – the last containing the grave of St Enda – the patron of the island. Nearby the ponies used to stand quietly and watch the Islander aircraft land on its sandy runway.

The precious hard-won fields on Inishmore are surrounded by drystone walls which are dismantled in regular spots to give access for the stock and then rebuilt with the animals inside. Along the coastal belt the island is criss-crossed with an interlacing pattern of grey walls where skylarks pour out their songs and corncrakes call. The notes and 'gowking' sounds of cuckoos are common as they are mobbed by their anxious meadow pipit victims. The chief settlement is Kilronan, where John Ridgway and Chay Blyth came ashore after their three months of rowing across the Atlantic. Behind the village is a wood, sheltered by the houses and the hill beyond, and among the sycamores and horse chestnuts you can hear the liquid notes of goldfinches and the moan of collared doves. Aran is Holy – 'Aran of the Saints' – and one is continually reminded of this.

From the hills of Inishmore on a clear day I could see the Dingle Peninsula while to the west lay trackless ocean all the way to Labrador. To the north the mountains of Connemara, where granite and peat are dominant, push up their mauve and rounded humps. Here is Joyce's country, and hills which make their own clouds of black, grey and white, vast peatlands which may serve a peat-fired power station, grazing lands for ponies and black cattle – Connemara at its wildest and loneliest. I visited the quarries from which came green stone and black flecked with white fossils. I walked across the bog at Clifden where Alcock and Brown landed after their historic flight; the site was indicated first by a huge stone monument like the tail fin of a Boeing 707 and nearer still a white stone like the nose cone of a rocket – a remote place where lay fragments of the radio station insulators whose masts had attracted the pioneer aviators.

But it is on Inishmore, where I saw more than forty different species of birds as well as many interesting plants, that I would like to end this chapter. The song of the Aran wren, of which I have a sound recording, is a permanent reminder of my visit – an encouragement to return and satisfy that yearning to go back which J. M. Synge called 'indescribably acute'.

> Oft the wind told me:
> Of Isles in the West.
> Islands of legend,
> Islands of rest.

4. The Whale-Backed Downs

I was first introduced to downland when I was scarcely ten years old. My earliest acquaintance with the rolling chalk hills was when I stayed with a freelance artist who lived near Westmeston in Sussex. The southern skyline from his bungalow was dominated by the great mass and scarps of Ditchling Beacon, which drew me irresistibly to an exploration of the grassy ramparts and chalk pits where I dug out fossil Cretaceous bivalves and prehistoric flint tools all covered with a white patina. Here I met rare orchids for the first time and watched red-backed shrikes along the northern downland slopes. In W.H. Hudson's time this was the home of countless meadow pipits, linnets, stonechats and whitethroats but the South Downs have changed since the great days of nature in downland. This line of chalk lies sandwiched between the present farmlands of the clayey Weald, which in the Venerable Bede's time was, in his words, covered with 'thick and inaccessible' forest, and the South Coast. These South Downs are not the only areas of chalk in England. Bold ranges of what Rudyard Kipling called 'our blunt, bow-headed whale-backed Downs' leap out from their centre on Salisbury Plain south into Dorset, north-east through Berkshire to the Chilterns, East Anglia, Lincolnshire and Yorkshire, and east into Hampshire before dividing into the North and South Downs.

Chalk downland has a very distinctive kind of scenery based on the chalk – a pure, soft, dazzlingly white limestone in the upper layers and greyer and less pure in the lower parts. Although chalk is a form of limestone its much softer nature distinguishes it from the older kinds of British limestone. It attracts certain kinds of plant and the presence of these calcicoles such as traveller's joy – our native clematis – and spindle tree provides a clear guide to the kind of countryside that one is passing through. As some kinds of flower, like foxgloves and broom, avoid chalk and limestone it is fairly easy to tell when one's journey is leaving chalk country, with its thin, dry soils. In Chapter 2 we looked at the limestone country of the Mendips; in this present chapter I want to consider the chalklands alone.

The chalk has been folded and tilted to form low rolling hills, streamless and riven by dry wooded valleys, sometimes with tree-fringed scarps that look down on vales of clay. The off-white cliffs of southern England's chalky bulwark against the Channel have become linked with the very name of England. A study of the chalk itself reveals that it is largely composed of the hard parts of very small floating plants called coccoliths and tiny fossil foraminifera. The Chalk Sea in which these organisms sank to the bottom was probably rather shallow and surrounded by low-lying land, perhaps desert, which was not being eroded away since the pure organic chalk is free of sediments that might have originated on land. Fossils such as sea-urchins are often common and nodules of black flint, made of silica and characterized by shell-like fractures, occur in layers in the chalk; flint was exploited by prehistoric man for tools and later for building materials and helped to form many of our southern pebbly beaches. Some of the chalk lies hidden under layers of boulder clay but, since in places it is over fifteen hundred feet thick, its deposition must have gone on for a very long period of time.

The North and South Downs both encompass the Weald with its beds of clay, and they meet to the west. The South Downs, with what Gilbert White called 'their gentle swellings and smooth fungus-like protuberances, their fluted sides and regular hollows and slopes', reach some sixty miles from Beachy Head and across Hampshire, where they merge into a great central area of chalk. Perhaps there is no better place to start a journey along the South Downs than from the great white precipice of Beachy Head itself – the medieval *beau chef* or Beautiful Head. Here is a strange world betwixt salt winds and inland breezes, which Richard Jefferies found 'wind-swept and washed with air; the billows of the atmosphere roll over it . . .' Samphire and sea beet cling to the steep chalk face and dwarf forms of slender-headed thistle, black knapweed, centaury and gentian grow in the narrow salt-blasted and foot-trodden belt of ground just inland from the cliff edge. Here in August I have braced myself to withstand a roaring warm south-westerly beating up the Channel to meet the tallest wall of chalk in Britain, where jackdaws tumbled and fluttered like tiny pieces of charred paper, herring gulls flapped and wailed and swifts scythed their sable way past the radar station. On one side lies the sea and on the other fields stretching away to the horizon, punctuated by coomb-like hollows and bottoms.

I made my way down to Birling Gap where grey-green combers were pounding the shingle beach and then up above the Seven Sisters – great white waves of chalk cut vertically as if by a gardener's edging tool – Haven Brow, Short Brow, Bran Point and four others. Behind lies a land of villages with greens and flint cottages and churches with squat towers and Sussex caps. This is the region which gave its name to a famous breed of sheep which once grazed the Downs in their thousands; it has been estimated that in 1801 Sussex boasted 342,000 sheep, most of which were on the Downs themselves. Windmills too added a special charm to the rolling countryside. Now large fields, some like prairies, have been ploughed across the Downs and corn has replaced the livestock.

Although the South Downs are no longer the romantic unfenced uplands that they once were, there are still some areas of more or less untouched turf with what W. H. Hudson called their 'close-bit thyme that smells like dawn in Paradise'. It is particularly on the northern slopes overlooking the Weald that this scented turf best survives. These are perhaps too steep to plough, are protected in reserves or by the National Trust or have remained unploughed by farmers at the request of the local authority or conservation body. Arable land has replaced grazing – a transition that started before the last war and on such a scale that it was decided to declare the South Downs an area of outstanding natural beauty and not a National Park. When the sheep kept down the scrub and maintained the turf – I can remember in the early 1930s seeing an old shepherd watching his flock and resting in the shade of a mushroom-shaped haystack chewed by the stock – the chalk downland was the rich home of many kinds of orchid, thyme, marjoram, vetches, rock rose, harebells and Hudson's 'great amethyst among gems' – the dwarf or stemless thistle – as well as chalk-hill blue and Adonis blue butterflies whose larvae fed on the horseshoe vetch. The Sussex Downs also show clear traces of Celtic fields while, older still, there are long barrows and causewayed camps and forts and ancient trackways which testify to early occupation by man. Today the Countryside Commission has established a modern South Downs Way which combines walking routes and bridle paths.

At the eastern end of the South Downs above the Seven Sisters and Cuckmere Haven are chalky hills of immense beauty and fascination – Combe with its ramparts and view across the Weald, Windover with the Long Man of Wilmington looking, in Rudyard Kipling's words, 'naked toward the shires' where willow warblers and white-throats sing in the wooded coomb, Firle with its 713-foot summit and the Caburn with its Iron Age fort where I found sweetbriar growing in profusion. Below lies Lewes – the capital of the Downs – with its red-roofed houses, partly restored castle and buildings whose walls have been constructed from blocks of black flint, knapped into almost perfect cubes. Westwards from Lewes is one of my favourite walks where the chalk runs in a fine imposing escarpment – Ditch-ling Beacon, Wolstonbury Hill, where on Ascension Day boys from Hurstpierpoint College sing 'Hail the day that sees Him rise', and the Devil's Dyke, with its rolling contours and beech-lined hollows. August is a fair month in which to climb Ditchling Beacon with its deep trench carved down the side; in the 1930s this was bare and grassy where the sheep grazed but now it is lined with scrub. I also walked from Brighton over the Downs but Patcham, then un-developed, has pushed its tentacles too far to the north.

One recent August I again walked up from the north through spires of ragwort and hogweed that now dominated the hill, but still amongst the grasses were clumps of knapweed and scattered dwarf thistles with their spiny rosettes – traps for unsuspecting holidaymakers! There was sprawling golden lady's bedstraw smell-ing of new-mown hay, bluish-lilac field scabious with pink anthers, and round-headed rampion – the Pride of Sussex – with blue birdcage-like flowers growing exactly where I had first seen it thirty-five years before. I also found ghostly skeletons of yellow rattle, eyebright with its yellow-spotted white flowers, pink storksbill and restharrow, purple thyme, nodding harebell and milkwort which the old herbalist Gerard thought 'fashioned like a little bird'. I could not find the little colony of haresfoot clover or the anti-clockwise climbing stems of the parasitic dodder that once grew there. Along my grassy path the rose-coloured blooms of bindweed straggled and ran riot among the swaths of wild privet, hawthorn and isolated wayfaring trees. Near the 813-foot summit of the hill bramble and hawthorn were proliferating and these were full of questing blue tits, while linnets twittered and sang overhead. I remembered with nostalgia how I had watched the sheep as a boy and feasted on the giant luscious blackberries that Ditchling had nurtured; from the top I could look down on Westmeston Church, where a pair of barn owls was nesting. Many years ago I used to see pyramidal, man, bee and musk orchids, hops, small nettles, musk and milk thistles and purple star-thistle but these now need more than a brief look to find. 'These blossoming places in the wilderness' Hudson, who so much loved downland, called them. It is one of the great pleasures in life to lie amongst the tiny flowers and bruised thyme and listen to the incessant almost birdless hum of the August afternoon on the Downs.

There are small circular ponds along the crest of the Downs – the 'dewponds' that once solaced the sheep that grazed nearby but which are now embraced by the cornlands. These ponds were 'sheep ponds' to the old shepherds – circular basins cut out of the chalk, lined with flints and rubble, then a layer of thatched straw and finally a coat of puddled clay brought up from the Weald. Rain and perhaps to a lesser extent mist kept the ponds filled up and the pond above Pyecombe dried up only once in fifty years. Kipling, the poet-chronicler of downland, knew about

> . . . the dew-pond on the height,
> Unfed, that never fails.

Some of the ponds were destroyed by wartime tank manoeuvres. Once considered to be of Saxon or even prehistoric origin the ponds are now thought to be no earlier than the eighteenth century. Amphibians, waterbugs, beetles and other aquatic insects live in the open ones but numbers have become rather choked with vegetation. I have seen warblers searching for insects in the vegetation, swallows and swifts swooping over the surface and many kinds of bird come to drink and bathe. Water is scarce in downland but the South Downs are interrupted by four broad river valleys, those of the Cuckmere, Ouse, Adur and Arun all flowing south into the English Channel. Beyond the River Adur a line of downs extends for over ten miles behind Worthing and Littlehampton. This is bare countryside cut into blocks by valleys and coombs with the wooded crown of Chanctonbury Ring – the best known landmark in the Sussex Downs with its beeches planted in 1760 by Charles Goring, who expressed the deep wish

> Oh! could I live to see thy top
> In all its beauty dressed.

I found rockrose, autumn gentian and devilsbit scabious common on Chanctonbury, On Cissbury Ring with its great hill fort and flint mines the skylarks sing an endless chain of song, the hares box and fight in spring and many butterflies wink across the grassy ramparts. From Beachy Head to the River Arun there are few wooded areas – I discount the 1500 acres of State Forest at Friston. The Arun marks off the downs and woodlands very clearly. As a schoolboy I described in my diary how I rowed up from the Black Rabbit near Arundel, with its imposing castle, almost to Amberley, observing

71

kingfishers and reed warblers on the way and the rare triangular bulrush. My diary records: 'This part of the Arun is almost tropical. The trees – beech, oak and yew – stretch down from Arundel Park on one side while on the other are great beds of reed and sedge. The trees are matted into one solid entity by masses of traveller's joy and convolvulus. At the foot of these curtains of vegetation grow colourful patches of purple loosestrife and ragwort. The feeling of being in a tropic clime is immediately removed by the vast rolling pit-scarred downland through which the Arun wends its peaceful and meandering course.' Hawker dragonflies and damselflies flitted along the river's edge, and on the way back I watched a roost of about twenty pied wagtails assembling in the reeds, to be followed shortly by over a hundred starlings. Arundel Park is a classic site for stinking hellebore, and gladdon – the roast-beef plant – and I also found spurge laurel in the same district.

West of Arundel the region of chalk is perhaps not quite so well known although the yew-woods at Kingley Vale and Butser Hill are not unknown to naturalists. Yew is perhaps the climax vegetation, especially on those slopes that are too steep to support beech trees. These yew-woods are dark and forbidding and the fluted columns of red-brown trunks hold up spreading canopies of dark green. At Butser the woodpigeon is the commonest bird in late May but there are also chaffinches, robins, blackbirds, goldcrests, wrens and other birds, while Kingley Vale at the same season has its chiffchaffs, willow warblers and turtle doves, but tits and warblers are not enamoured of these shady woods. Hairy violet makes these woodland fringes blue with colour. In the summer days grasshoppers chirp, flies and humblebees buzz and drone over the turf around the yew-woods and wolf spiders stalk their prey over the warm, chalky soil.

From Sussex the chalk widens out into the great mass which extends to Salisbury Plain. North Hampshire was familiar ground to Gilbert White and Selborne Hanger – the beechwood above the village – is just as attractive as it was in his day. Twice recently I climbed up the zigzag path built by him and his brother in the chalk, to walk among the trees and to listen to those three leaf-warblers – chiffchaff, wood and willow warbler – which gave him much pleasure and difficulty but which he was the first to separate. Salisbury Plain itself is a great undulating piece of countryside broken by occasional scraps of woodland, belts of thin scrub, ancient earthworks and areas which the military has closed to the general public. Tall broomrape – a parasite on greater knapweed – raises its characteristic spikes all over the chalk Plain. In some places arable, woodland, grass and thorn or juniper scrub grow side by side. Juniper is now spreading as a result of the disappearance of rabbits caused by myxomatosis. On the grasslands lichens and chalk-loving grasses and vetches produce a mat rather like that of Breckland in East Anglia. Here skylarks are the commonest birds but yellowhammers, mistle thrushes, stonechats and hedgesparrows favour the clumps of thorn and occasional elder. Lapwings and stone curlews nest within the protected military areas and buzzards and hobbies often hunt over the region. I once spent a week making tape-recordings of a pair of hobbies whose nest was in a classic situation – in an old crow's nest high in a pine tree in a clump on high ground and from which there was a wide view across the open country that stretched away to the horizon. This rare and handsome little falcon has a special food presentation ceremony; I watched the male call the female off the nest so that she could take his offering – perhaps a skylark, a yellowhammer or a large grasshopper.

In Dorset the chalk belt may be less interesting botanically, particularly in some areas such as Dorchester but the Pentridge Hills are rather better. However, one day to the steady musical chatter of a combine harvester I began to climb the long grassy slope that reached to a Dorset hill-top – a great bank of chalk some thousand yards in length. In the words of Thomas Hardy, it has 'an obtrusive personality that compels the senses to regard it and consider'. This was Maiden Castle – shaped in three different eras – the Neolithic and early Bronze Age, the early Iron Age and the early Roman period, spanning the years from 2000 BC to AD 70. Somewhere about the year AD 43 or 44 the Second Augustan Legion under the command of the future emperor Vespasian stormed the formidable defences and cut down the defenders, one of whom lies at peace in Dorchester Museum with a ballista arrowhead in his spine. At the summit beyond the western rampart fat sheep and Friesian cattle were grazing, and I watched wall and blue butterflies dancing in the sun. Meadow pipits and yellow wagtails stalked back and forth among the cloven hoofs, and starlings leapt up to pick a fly from the muzzle of a grazing cow. There were rows of thistles, their seeds adrift on the wind, and among these flittered goldfinches and on one dry stalk – bold and self-assertive – stood a migrant Greenland wheatear. Kestrels were hovering over the grassy earthworks, where rough flint flakes on the exposed chalk betrayed the presence of flint-knapping man. Yarrow, scabious, eyebright and gentian were still in flower. A cock linnet sang some phrases from a post near the Roman temple, and its inconsequential notes seemed only too well to reflect man's evanescent association with the great hill. After all, by

AD 70 the population of Maiden Castle had been moved down to Dorchester.

Chalk appears even further west on the Devon coast, and the flora around Branscombe includes such interesting plants as rock sea lavender, rock spurry, samphire and Portland spurge. The white chalk also forms strips behind Lulworth Cove, from the Hills to Old Harry Rocks near Studland. Corfe Castle stands on a chalk mound, providing a tall romantic ruin for the fortress where King John kept his treasury and State prison; on the mound I have found borage and the shining cranesbill in flower, while jackdaws gather on the ancient walls. I have also seen street pigeons and once a stock dove fly out from crevices in the grey ramparts. On the downs near Studland henbane grows – that greeny-buff hairy and poisonous plant with its creamy purple-veined bells of flowers.

The chalk strip near Swanage looks eastwards towards the Isle of Wight, whose backbone is a narrow ridge of chalk with almost vertical strata running some twenty miles east to west across the island. This chalk 'hog's back' was the result of Alpine folding in Miocene times and lies between belts of sand and clay which at Alum Bay occur in upright bands. In the west the chalk formed the Needles and the high downs at Freshwater; in the east it rises from the sea near Sandown. On the high downland turf many of the chalk-loving plants that we met on the South Downs and at Beachy Head occur, with pyramidal orchids and bastard toadflax, but there are some notable absentees. Strangely enough the flora of the Ventnor chalk is somewhat different again and includes rare cuckoo pint and the attractive purple cow-wheat. There is a great attraction about the chalk downs of the Isle of Wight which I have seen both in winter and the middle of summer. I have walked the walls of Carisbrooke Castle just as the unfortunate Charles I did during his captivity, and listened to the voices of pied wagtail, collared dove, starling, wren, hedgesparrow, blue tit and mistle thrush. On some days I have seen the island swept by gales, hail and torrential rain, with lapwings gathering on the grasslands and parties of curlews and black-tailed godwits flying low over the downs on their way to Newtown marshes to join the Brent and Canada geese, golden and grey plover, turnstones, greenshanks and redshanks, dunlin and ringed plover that I have watched on the mudflats and along the creeks of this wildlife reserve.

To me the Isle of Wight is in some ways a miniature England yet with its own special appeal. Inland are narrow winding lanes of great charm, bordered with hollies, holm oaks, laurels, tamarisks, pines and cupressus, small copses full of wind-blasted oaks and hazel where long-tailed tits and treecreepers forage, seaside towns with their Regency and Victorian villas, stone and brick cottages and farms, a Royal home whose grounds are full of woodland birds and in the distance that range of chalk downs which rise to over 700 feet, often into a grey cloud base but which give to the landscape a strange, wide dimension.

After our island digression we must return to North Wiltshire, with its earthworks and often interesting chalk flora. I have clambered up Silbury Hill, just off the Bath road, scaling 130 feet of that great grass-covered plum pudding of chalk to the whisper of yellow-hammer songs and the rusty jingle of corn buntings. Round-headed rampion has been known here for nearly three centuries and is one of the gems of Silbury. Avebury, with its great earth rings and standing stones, is nearby, once standing among many elms on the edge of the Kennet water-meadows and graced with the radiant blue of meadow cranesbill. I have seen wheatears sitting on the stones of the Kennet Avenue which leads to the entrance to Avebury and a weasel sprinting from one piece of grass cover to the next. The Marlborough Downs are littered with the Sarsen stones, or Grey Wethers, some of which were used to construct the Avebury circle, and they too have a rich chalkland flora. The great central mass of chalk on Salisbury Plain and in Hampshire branches to the north-east to the Chilterns and east to the North Downs. Before returning to Avebury in order to follow the ancient Icknield Way along the Berkshire Downs and the Chiltern Hills it is necessary to see the North Downs which spread in a hundred-mile ridge from Dover across Kent and Surrey. These downlands are close to London and so are well known to many visitors from the metropolis and abroad. In the most popular places they are subject to considerable erosion and damage from too many passing human feet. On clear days I can see the North Downs across the London Basin from Dollis Hill in north-west London where I live.

The chalk ridge in Kent is the home of some of our finest and most local orchids – lady, early and late spider, man, fly, fragrant, pyramidal and autumn lady's tresses – and such calcicole specialities as Kentish milkwort. There are also woods of beech, yew and box. West Kent rejoices in hairy mallow, ground pine and meadow clary, and the Pilgrim's Way gives ready access to many of the best chalkland sites. Indeed Kent has the richest limestone flora of any county in the British Isles. The North Downs in Surrey are not a true continuation of those in Kent and lack some of the Kentish floral glories. They have a gradual northern dip slope with steep-sided dry valleys and a very steep southward-facing scarp with chalk close to the surface.

75

Box Hill, which I know well, is capped in places with a layer of 'clay-with-flints' and although beeches do occur, oak and yew are also common and well developed. I have passed many happy hours walking up the steep chalky slope, where at the weekend thousands of Londoners toil up its grassy hill to admire the view and perhaps the downland flowers. The National Trust owns the hill and I know quiet valleys and woods where, away from the viewpoints and tea-places, there are spots to roam disturbed only by the passage of jets on their way to and from London Airport. Early morning is a special time to walk under the tall beeches and among the graceful, sombre yews. There is a dewy sparkle on the grass and scores of grey squirrels forage among the beechmast or sit in the trees and petulantly complain with their whining, scolding voices. Magpies rattle and jays scream in the canopies of the trees and warblers sing in the lower growth. There are badger setts where one can see the animals playing and feeding or bringing in piles of new bedding clamped firmly between their front legs and their lower jaws as they move backwards towards the burrows, stopping every now and then to listen.

On the lower slopes of Box Hill is Juniper Hall – one of the centres of the Field Studies Council, from which I have often explored the downs. There is a dense grove of box from which the hill gets its name as well as oaks, beeches, yews and whitebeams. Arboreal tits, willow warblers, woodpeckers and treecreepers are common. Shrubs of hawthorn, dogwood, wayfaring tree and privet – the latter sometimes with its complement of hawkmoth caterpillars – grow up the slopes. Roses, traveller's joy and spindle tree are there to delight the visitor as well. In May the grassy banks are bright with violets and milkworts, and soon a succession follows of man and fragrant orchids, pyramidal and bee orchids, followed by rock rose, kidney and horseshoe vetch, birdsfoot trefoil, squinancywort and yellowwort. By late summer wild carrot and parsnip, small scabious and dwarf thistle, thyme and marjoram, eyebright and clustered bellflower and autumn gentian have joined the cavalcade of downland flowers. Some will go on flowering into autumn. In the beechwoods dog's mercury, cuckoo pint and wood spurge are followed by the large white helleborine and bird's nest orchids. A dozen species of grasshopper and bush cricket have been recorded on Box Hill as well as chalk hill blue, skippers, ringlet, green hairstreak and dark green fritillary among the butterflies.

The oldest trackway in Britain is the Icknield Way which runs across country from Avebury and the rolling downs of Wiltshire to the Berkshire Downs, across the River Thames to the Chilterns and the dry brecklands of East Anglia. This was the flint way along which the prehistoric flint-knappers of Grime's Graves were linked by traders with the men of Wiltshire and the monuments of Avebury and Stonehenge. This route is a marvellous way to see the chalk hills of Berkshire and the Chilterns and I have walked it a number of times. The first section runs from Hackpen Hill with its association with Hilaire Belloc and the leaping horse to Liddington Castle – an Iron Age camp and a favourite haunt of Richard Jefferies. Ahead lies the White Horse of Uffington that has sprawled and cantered across this green brow of hill – a chalk landmark – for two thousand years. In July the approach road and the embankments are a wild garden rich in colour, especially the blues and purples of high summer. There is the dark blue of nettle-leaved bellflower, the violet-tinted blue of meadow cranesbill yet again, the pale lilac blue of scabious and creeping thistle, the glowing pinkish purple of rosebay and the royal purple of knapweed. Ragwort, St John's wort and agrimony push up golden spikes towards the sun. Along the south-facing bank of Uffington Castle I have watched the flies and humblebees move fitfully from blossom to fresh flower and back again. Horseshoe vetch and milkwort are common along the Berkshire Ridgeway section of the track but the large autumn gentian seems to be confined to one earthwork. An autumn visit to White Horse Hill can be a rewarding one, as chaffinches, starlings and skylarks with smaller numbers of other finches, pipits, wagtails and thrushes move in the early morning west along the scarp of the Downs. There is also the mystery of Wayland Smith's Cave – a megalithic long barrow just by the Ridgeway where green woodpeckers yaffle – and the Blowing Stone at Kingston Lisle, which I made a detour to see and blow; I produced a poor sound unlike the lady in the cottage nearby whose deep horn-like note echoed round the village – a trumpet by which, according to legend, King Alfred summoned his armies. The stone – a red sarsen – once stood on the Ridgeway.

The River Thames acts as a natural boundary between two different kinds of chalk country – the open treeless pastures of the Berkshire Downs and Ridgeway, and across the ancient ford to the north-east the heavily wooded Chilterns. Vast wooded slopes, some probably planted, tower above the river and look down on flat acres of pasture meadows. There are long reaches where after the hay crop and the grasses have been cut down the sedges and stately flowers of summer assume new stature and rise taller in their competition for the life-giving sun. The banks are clothed with hemp agrimony and meadowsweet, purple loosestrife and great willow herb, and the river is alive with moorhens and their querulous young, coot, mallard, mute swans, Canada geese, yellow wagtails and sedge warblers. The air shimmers with the wings of dragonflies and damselflies, dace and chub rise to the surface to suck down the flies that have fallen on the surface, while below the surface nymphs and other voracious larvae live out their intermediate lives as freshwater assassins.

The Icknield Way crosses the Thames near Goring and Streatley and climbs up on to the high ground of the chalk Chiltern Hills, but in some places the chalk is overlaid with clay-with-flints, brick earths and gravels. Magnificent beechwoods – the source of timber for the local furniture industry in High Wycombe – flourish on the well-drained soils. A steep scarp with just a thin layer of soil, and rich in orchids and other downland plants, extends from Whitchurch to Princes Risborough, Wendover and Tring. On Swyncombe Down I have seen wild candytuft, large autumn gentian, dropwort and pale toadflax growing in the summer turf, while in autumn before the last war the stone curlews used to gather there twenty or thirty at a time. The beechwoods and their slopes are the home of such rare plants as the military and ghost orchids as well as monkey, lady, man and green-flowered orchids. The clay on the Chilterns actually permits acid-loving plants such as heather and broom to grow close to rock roses! Among the Chiltern butterflies are dark green fritillaries, green hairstreaks, small and silver-studded blues, chalk hill blues, brown argus and marbled whites.

One Easter morning I watched the dawn rise on the Chiltern scarp. The air was pleasantly mild, the ground dank and slippery with dew and little isolated wreaths of mist were clinging to some of the hollows. From an open space on the hillside came the liquid notes of a woodlark in sequences of short and slightly varied combination. The song lacks the passion and vehemence of the related skylark but it is enhanced with fluent and melancholy cadences of limpid beauty. Blackbirds and robins soon began to sing and then the double call of a cuckoo and the brief incomplete notes of a redstart joined the chorus of birdsong. By now a pale blue-grey light was beginning to illumine the sky behind the scarp. Patchy clouds hung overhead as the countryside began to stir. Blackcaps and willow warblers were everywhere and the short ditties of yellowhammers and chaffinches came along to round off the morning. Beyond Bledlowe Cross, carved in the white chalk on Wain Hill and situated in one of the finest beauty spots in Buckinghamshire, I found a narrow sheltered lane running up a spur of the Chilterns across the Icknield Way and bordered by beeches and wayfaring trees. Here a cirl bunting – a rare relation of the yellowhammer – was singing his short song, full of metallic repetitions and suggestive of the rattle of the lesser whitethroat which also comes to the high thorn brakes along these hills. At this point in the Chilterns there are two Icknield Ways – the Upper which is perhaps the line of the winter route and the Lower the summer track of drier weather. In the well-drained upper chalk of these hills, where the woodlands border the open country, badgers often excavate their setts, showing a clear preference for this over the lower chalk and often sharing their underground homes with foxes.

To the east of Tring the handsome Chiltern beechwoods begin to run out and the open downland boasts purple milk vetch and field fleawort. The best of the chalk flowers are to be found on the hill slopes that face north and include such well-known spots as Ivinghoe Beacon, where model aircraft compete with the many hovering kestrels, and Dunstable Downs, which were the scene of great autumn passages of skylarks in the eighteenth century, according to the naturalist Thomas Pennant; I have observed passages of larks in autumn as well as movements of redwings and fieldfares along the Downs.

To the east of Hitchin there is no longer a scarp and much of the land is cultivated, so that plants of chalk can only be found on any sharp slopes or ancient earthworks which the plough could not reach. The coming of the Enclosure Acts was to result in the progressive destruction of chalkland plants in Cambridgeshire although some have managed to survive as weeds of cultivation. Therfield Heath lies to the side of the Icknield Way and more than four hundred acres of its thin chalk grassland, practically untouched by man, are now a local Nature Reserve. There is a fine chalk flora with cowslips, wayfaring trees, rock roses, birdsfoot trefoil, lady's bedstraw, wild carrot, burnet saxifrage, milkwort, self-heal and clustered bellflower; there are also the rare purple milk-vetch and bastard toadflax and four orchid species, one of which was introduced in 1965. I have seen nearly a dozen kinds of butterfly on Therfield Heath including the Essex skipper. Among the bird species tree and meadow pipits, skylarks, cuckoos, tree sparrows, warblers and partridges provide testimony to the variety of habitat on the reserve.

The chalk continues to run north-eastwards and on certain dry grasslands and ancient earthworks the magnificent pasque flower still survives. There are rare poppies and larkspur, and the tuberous thistle still clings to an old trackway. The Gog Magog Hills, where other tracks with wide grassy verges run through fields of corn and between clumps of pines and broad-leaved trees, are still home for grape hyacinths and stars-of-Bethlehem, perennial flax and mountain stone parsley. Redpolls twitter and wheeze and many years ago red-backed shrikes could be seen here as well. The Icknield Way, sometimes old track, sometimes modern road, takes us all the way from Avebury into Breckland – that steppe-like region lying in the triangle of land formed by the three towns of Swaffham, Newmarket and Bury St Edmunds. Here a layer of sand, that once drifted, in John Evelyn's words, 'like the Sands in the Deserts of Lybia and quite overwhelmed some gentlemen's estates', overlies the chalk. This loose top coat has probably been here since the Ice Age and now forms desolations of heathland where pine, birch, heather and bracken have taken over. There are thousands of acres of State Forest made up largely of Scots and Corsican pine. Here I have watched red and roe deer grazing in the clearings and red squirrels shredding cones while crossbills 'jip' overhead – exciting, exotic birds which always bring a thrill of recognition.

In places the sand is thin and the chalk is quite close to the surface. Pits will often show the nearness to the surface of the white chalk, and the mines at Grime's Graves show how the best flint – the floorstone – was excavated by ancient man at a depth of around thirty feet, using red deer antler picks and shovelling the chalk with the shoulder blades of oxen. I found it very moving to stare at the soot marks on the tunnel roofs left by the miners' lamps. I once demonstrated for a BBC television film the method of flint tool-making used by the miners, sitting not far from one of the pit shafts. Flint was also the local material for building houses, churches and walls and I have seen walls constructed of chalk as well. I have often travelled the route from Thetford to Brandon, where the art of flint-knapping was still carried on. Redpolls dipped across the road and I saw a roe deer clear a six-foot fence from a standing jump. Tawny owls would often sit in daylight on posts along the road and woodcock would go 'roding' between the tree trunks. The roadside was often gay with the purple of rosebay, the glowing blue of viper's bugloss and the golds of Aaron's rod, ragwort and lady's bedstraw. In Brandon parties of swifts would fly over the 'Flintknappers' Arms', where the incisive tapping of the flint knappers at work could be heard over the sound of the traffic. Later the work was continued at the home of Herbert Edwards, knapper-in-chief, on the outskirts of the town. Four streams flow through Breckland – the Wissey in the north, the Little Ouse and Thet in the middle and the Lark in the south, all winding their way westward to the Fens and carrying very calcareous water with them. Breckland itself burgeons with flowers and where the chalk is near the surface I have been delighted with views of maiden pinks, spiked speedwells, purple milk-vetch and burnt orchid, the latter growing not very far from a stone curlew's nest.

There are not many exposures of the chalk in Suffolk and Norfolk or across the Wash in Lincolnshire, where most of the formation is overlain with glacial drift. However, the Lincolnshire Wolds with their steep scarps, narrow valleys and lonely, treeless sweeps of country once devoted to sheep are truly chalk hills. The wolds bear some resemblance to the South Downs but have been described as 'burlier' and less elegant. They run for some forty-five miles and their scenery is more northern where oak and ash replace the southern beeches but much of the area is under cultivation. Winding roads with quickset hedges and wide verges from which isolated trees raise their boughs link the brick-built farms and houses. On the uplands there is little water and most Wold settlements are in the valleys. The Western Wolds are picturesque and high, dominated in one place by radar reflectors, clothed with vast pasturelands and arable fields where skylarks and lapwings nest and dotted with occasional woods of oak, ash and wych elm sometimes ornamented with guelder rose. I have walked in autumn through a mixed wood on the edge of the Wolds and, as I made my way round a group of thorns, a party of half a dozen bullfinches flashed red and white as they flew away from me. There was a nomadic party of great, blue, marsh and long-tailed tits working their way quietly and without fuss through the tree canopies. In a tall yew tree six goldcrests were calling with high-pitched 'zee-zee-zees' as they climbed and swung beneath the branches. A stoat ran across a ride carrying a shrew and dived rapidly into a clump of undergrowth. Hares tooks refuge in the scrub cover where in the evening greenfinches and chaffinches came to roost, while parties of rooks and jackdaws assembled in a line of tall beech trees. The high Wolds I knew especially well during the last war when I used to cycle winter and summer from my bomber station at Wickenby to visit my wife, who was a WAAF officer at RAF Binbrook when the Royal Australian Air Force was there. In the summer there were corn buntings, yellowhammers and linnets to join the countless larks that gave me encouragement on the way, and in winter short-eared and barn owls hunting along the hedges and over the airfield, and even an occasional skein of pink-footed geese from the Wash.

The most northerly exposure of the chalk in England and one which brings us to the end of this chapter is in eastern Yorkshire. It begins just north of the Humber and stretches to the Vale of Pickering. Again there is a considerable amount of glacial deposit on the surface and much of the land is under cultivation. There are, however, the magnificent white sea precipices at Bempton and Flamborough Head. J. E. Lousley, who was an authority on the flora of the chalk and limestone in Britain wrote: 'in general, Yorkshire

81

chalk lacks many of the attractive southern plants and has very few additional northerners by way of compensation'. You can still find rock rose, birdsfoot trefoil, salad burnet, marjoram and clustered bellflower, especially on grassy sidings and steep slopes too acute to plough such as those that occur around the typical Yorkshire Wold villages of Millington and Heygate. There are huge and open landscapes criss-crossed with roads that to me suggest loneliness and isolation. In the north the old uncultivated lands were enclosed to become first-class farming land. Folds of chalk in the Wolds are separated from each other by dry valleys, often with a fringe of thorns. The 'Wolds Way' is a good route for seeking the chalk landscapes of the East Riding although the official track needs to be used by walkers. There are prosperous farms nestling inside their shelter belts, villages with broad streets and ponds, some of which may be cut off by winter snow, ancient Bronze Age cemeteries, Roman villas, Norman churches, more than a hundred deserted medieval villages, chalk quarries, beechwoods and a typical range of woodland and farmland birds.

For the naturalist perhaps the greatest attraction of the East Riding lies in the white precipices that mark the end of England's chalk – Speeton Cliffs, Buckton Cliffs, Bempton Cliffs and Flamborough Head. Bempton Cliffs are partly private, partly owned by the Royal Society for the Protection of Birds. Here is an important colony of seabirds – herring gulls and kittiwakes, guillemots, razorbills and puffins, fulmars and shags and perhaps from fifty to a hundred and fifty pairs of gannets – the most southerly colony on Britain's east coast. Bird visitors are carefully logged at Bempton and many include warblers, swallows and cuckoos, even shore larks and a bee-eater, while from the cliffs it is possible to watch migrating skuas, including the scarce long-tailed, and terns. High winds sometimes affect the breeding success of the kittiwakes and oil spillages that of the auks. Many gannets at Bempton have died lashed to their nests by abandoned lengths of nylon fishing line unwittingly brought in by the birds as nesting material. The cliffs at Bempton were exploited in the past by professional egg-gatherers dangling in space at the end of ropes in front of the cliff-face which has claimed some birdwatchers in recent years as well. These great walls of white chalk are one of Britain's greatest natural spectacles – dramatic, noisome and in summer alive with the calls of seabirds – an appropriate complement and end to the chalk downlands which we first met on the South Downs and at Beachy Head. The chalk with its often beautiful and rare plants and animals is part of our natural heritage and deserving of a very high place in our conservation planning.

5. A Wind on The Heath

In Southern England the chalk downlands with which we were concerned in the last chapter sometimes enclose broad shallow basins. The best known of these depressions are the London and the Hampshire basins, filled with sands, gravels and clays, and encompassed by the chalk Chilterns and the Hampshire and North Downs. The heathlands of Dorset and Hampshire can be found in the Hampshire saucer, which is partially flooded by the waters of Poole Harbour. To the north the chalk downs of Cranborne Chase and to the south the grassy flats of the Isle of Purbeck provide the rim to this natural basin. It may be that a large river discharged its gathered materials into a wide estuary or shallow sea that invaded the older chalk surface and left behind some of Britain's more recent 'rocks' – the Eocene sands, clays and gravels. The rather acid soils that evolved from these deposits were to produce somewhat barren areas, often as flat-topped plains, with gorse, heather, tough resistant grasses like purple moor-grass and bristle bent, and occasional self-sown birch and pine – an unusual landscape in the middle of rich agricultural lowlands and repeated also on the Wealden sands and Lower Greensand of the Cretaceous period.

There are many heaths in Surrey, and the Bagshot Sands account for the heathy nature of the summits of two of London's open spaces – Hampstead, where nearly a hundred different species of bird have been seen in a recent year with more than a third actually nesting, and Highgate. It is only in the more fertile eastern part of the Hampshire Basin that the soils and sandy loams of the New Forest support areas of woodland, especially those of beech and oak with some holly, yew and thorn brakes. These woods tend to cover the slopes of the barren heathy plateaux; they may be founded on layers

83

of white pipe clay containing marine fossils and shells, which were exploited from Roman times in the pottery industry. There are also many low-lying areas of poorly drained marshland where willow, alder, heath, bracken, bog-moss and cotton-grass flourish in the rather waterlogged ground, and the fronds, branches and white seed-heads nod in the wind on the heath.

The natural woods of beech, oak, holly and blackthorn of the New Forest offer us enchanting and welcome woodland glades. Here with the surrounding heathy waste was one of the Royal Forests mentioned in the Domesday Book – a royal hunting ground where William Rufus met his death and where today the Commoners' animals, the cattle and ponies, keep open those delightful glades which form the characteristic New Forest landscape, being 'the architects of its scenery'. The Forestry Commissioners generally control the Forest but the duty of overseeing the common rights and Commoners' animals rests with the Court of Verderers and their officers, the Agisters. The Court meets at intervals of around six weeks in the Verderers' Hall in the Queen's House in Lyndhurst. There are many enclosures in the New Forest of alien conifers and hardwoods planted by the Forestry Commission and two collections of rare trees open to the public. The Commission controls the numbers of deer for the health and quality of the stock and to avoid too much damage to forestry or agriculture. Before culling many deer starved to death because of their over-exploitation of the food reserves in winter-time. Under the New Forest by-laws wildlife in the new forest is protected, and no digging up of plants, hunting, fishing or catching insects even for scientific purposes can be carried out without an approach to the Deputy Surveyor to seek permission.

I often drive through the New Forest along the A31 (T) road from Cadnam and am always intrigued to see parties of people with their cars parked in a noisy lay-by, picnicking within a few feet of the roaring wheels and diesel fumes, when a short detour of a few hundred yards could have taken them along quiet leafy lanes where the deer cross the road and the glades are full of birdsong. In spring the woodland floor is carpeted with wood anemones, wood sorrel, primroses and bluebells while the observant visitor may spot golden saxifrage. I have seen early purple, butterfly, common twayblade and broad-leaved helleborine orchids. There is bastard balm, butcher's broom, columbine, lungwort and my greatest reward has been the finding of the wild gladiolus blushing pink under a bracken frond. The broad-leaved woods of the Forest are full of crows, jays, the three kinds of woodpecker, nuthatches, tawny owls, woodcock and buzzards which take their toll of medium-sized birds, rabbits and woodmice, frogs and occasional reptiles. In the conifer plantations the commonest birds are goldcrests and coal tits but there are sometimes long-eared owls and crossbills as well. Within the Forest limits also live native British species of amphibian and reptile including, of course, the rare smooth snake, and natterjack toad. The ponds contain crayfish, sticklebacks, eels and lampreys as well as leeches, water spiders, water boatmen and beetles, snails and whirligig beetles and there are small brown trout in the streams.

I have walked through the Forest in every month of the year. In May we enter into the full luxuriance of summer when many of the trees and shrubs are clothed in their first delicate green and the great dawn chorus of the birds is unfolded like a vast tapestry of sound. Although only two hours by train from London the New Forest still retains its secret glades, moors and swamplands little changed since the time of William the Conqueror. One May morning – about four o'clock – I stood in a forest clearing with a tape recorder waiting for the dawn. The air was cool and clear under a steely moon, the dew dripped from the trees and saturated the grass, and a woodcock went 'roding' between the oaks. A blackbird uncertainly and rather faintly began his first phrases from the top of an ancient stag-headed oak. Soon other birds began to tune up – robins and song thrushes – and a tawny owl, incisive and insistent, began to 'kiwik'. As a pale blue light widened across the forest top the chorus of song began to swell. For me this was pure delight, unlike the experience of Shakespeare's insomniac who yet does lie

> Sleepless! and soon the small birds' melodies
> Must bear, first uttered from my orchard trees:
> And the first cuckoo's melancholy cry!

And as the cuckoo began to call there was sufficient light to glimpse the little group of ponies – a grey stallion and three chestnut mares – cropping the damp grass by the forest edge. How different from other mornings when I watched the ponies unhappily shaking the water from their backs as the rain poured down into a dank and misty landscape. Some young calves, black and woolly, were crossing the road and several fallow bucks, some still antlered, others newly deprived of their glory, were browsing restlessly forward into the clearing. The sun rose quickly, casting long, low shafts of misty ethereal light between the oaks. A song thrush declaimed his joyous phrases as six o'clock sounded from the spire of Lyndhurst Church. Starlings began to search for worms in the grass and a green woodpecker to probe an ants' nest with his long sticky tongue. Suddenly a red deer stag appeared and majestically crossed a heathy

85

clearing before disappearing into the shelter of the trees. The New Forest is famous for its warblers and I heard no fewer than ten different species. One of the commonest was the whitethroat, whose short jerky phrases seemed to come from every bramble brake. Blackcaps were abundant, singing their clear fluent songs from the canopies of the trees, while the sustained contralto notes of garden warblers could be heard from the denser cover of vegetation under the trees. It was, however, the wood warblers among their commoner relations, the chiffchaffs and willow warblers, which brought me the greatest pleasure. It was possible to hear eight birds in song at once. Short shivering trills were succeeded by plaintive plangent notes sometimes to the accompaniment of a redstart in the background. I once recorded some wood warblers singing in a beech grove as the drops of rain fell off the green shiny leaves.

When the longest day is past the bird chorus in the Forest begins to die away, leaving the stage very much to the doves, finches and buntings and a few other kinds of bird. One July afternoon I kept a solitary vigil balanced in a tree from whose fork I could peer out over the tall green parasols of the bracken fronds towards a badgers' sett. A low shaft of sunlight pierced the rich green canopy of the tree, enhancing the scene with a strange beauty and striking across a grassy ride that led down to a belt of conifers. In the sun a tall marsh thistle raised its crown of small purple flowerheads – the landing platforms for high brown and silver-washed fritillaries, humblebees, hoverflies and once – and what a moment that was! – a rare white admiral butterfly. From the wood on a distant crest came the drawn-out plaintive call of a buzzard as the bird left the trees and began to soar on broad pinions above the green billows of the Forest. It had rained earlier in the day and the running water had washed the ground, exposing some scrapers and flakes once fashioned by Bronze Age Man and a number of cracked pot-boilers. The midges came out in countless thousands but liberal application of dimethylphthalate kept them at bay. From the conifers the thin delicate song of a goldcrest drifted over to me on the lightest of summer breezes. Suddenly and without warning a fallow doe, chestnut-red and spotted, leapt out from behind a grassy bank into the centre of the ride below me. A slight shift in the wind – perhaps a whiff of my scent – and the doe gave a crisp, resonant bark of warning to her fawn lying up in the bracken, and she was gone! Two jays came dipping low over the trees and screamed at the sudden sound. Then a burst of angry chittering calls from some robins and wrens gave notice that they had stumbled upon a tawny owl which began to hoot in strained, offended tones.

The light now began to fail and a roe deer crept out from the bracken, moving silently away from my tree with exaggerated steps as if walking on broken glass and with her big ears making her look like a strange long-legged fox. A woodcock began to flight above the oaks, calling with high-pitched notes, interspersed with coarse amphibian-like croaks – 'quarnk-quarnk-quarnk-tit-ic'. As the sun finally dropped behind the horizon, a young tawny owl began its insistent hunger call and a nightjar, lying close along a branch like a giant froghopper, launched into his throbbing, rattling song. I listened – unsuccessfully as it turned out – for the rare New Forest cicada, Britain's only representative of this large family. As the light almost totally disappeared, my watch was rewarded by the appearance of a sow badger and three cubs. They scented and tested the air and then like playful, uninhibited puppies they embarked on a game of 'tig'. Food was their next concern and the family ran off along an ancient badger track that led straight to a mossy clearing, where beetles and bluebell roots would provide them with a favourite meal.

August in the New Forest means plenty of cars, caravans and campers but in 1963 Bank Holiday weekend rains washed the camp-sites away. I have often wandered the Forest in late summer and walked through glades of oak where I could almost reconstruct the ancient scenery of our lost primeval forests in spite of the missing tangle of fallen trees and deeply mouldered boles that once made half of Britain an impassable hazard to the traveller. Wrens go singing into August, and this tiny bird with the high passions of its song encircles the globe. Above the bracken tops I have spied the tossing antlers of the fallow deer which, though not indigenous, have provided centuries of sport and food in the Royal Forest. I once stalked a herd of twenty-two out in the open and was able to watch and film the bucks, does and fawns, all quite unaware of my presence, their black tails with white undersides busily flicking away the flies. Amongst them walked eleven carrion crows catching grasshoppers disturbed by the deer as a magpie sat watching from a pine.

The woods and copses of the New Forest foster an exceptional number of dependent species – a multitude of insects and, of course, grey squirrels that whine away from the August oaks, deer, badgers and foxes, stoats and weasels, mice and voles as well as nuthatches, treecreepers, woodpeckers, stock doves, and tits, while the slurred, unformed calls of juvenile jackdaws can be heard in forest and town alike. The sharp axes of our ancestors made great deciduous forests rare in Britain and so the New Forest has a special interest for the naturalist. The branches of the trees are foraging grounds for vast numbers of moth, butterfly and other insect larvae. I have seen stag and minotaur beetles as well as wall, large white, gatekeeper, and large.skipper butterflies, while huge fierce Tabanid flies or 'stouts', especially *Tabanus sudeticus*, may make life a misery for the walker in the forest.

Without any doubt one of the great attractions of the New Forest are the deer although many visitors drive through without knowing that they are there or only catching a glimpse of a disappearing black and white rump under the trees. There are four species – red, fallow, roe and Sika. A curious October sound in the New Forest is a strange rising and falling whistle with a concluding grunt. This is the rutting or mating call of the introduced Sika stag. This whistle, which is quite unlike the call of any other British mammal, is given mostly at dusk and in the early morning. The sound is produced three or four times and is followed by a long period of silence lasting perhaps an hour or more; this makes it very difficult to locate a rutting Sika stag, which is a shy animal anyway. It was certainly one of the most difficult wildlife sounds that I ever recorded but one evening a fine dark brown stag moved out into a ride to browse and ten minutes later gave the call. Coming as it does usually from concealed animals and after a long period of silence I find the whistle quite startling and it always makes the hairs on my neck stand up. The rut of the Sika deer lasts from September to November but there is still a great deal that we do not know about this rather shy deer.

The other more familiar deer of the New Forest – the fallow – rut in October. Once I took my recording gear into a wood deep in the heart of the Forest with a shrub layer of bramble and holly. Two hours before dusk I hid myself near a fence on the edge of the woodland where a fallow buck had its rutting area under some tall beech trees. A fine drizzle was falling and there was not a breath of wind to move the bronze and yellow leaves of the trees. It was still warm enough to encourage a great tit into a burst of autumn song, and as the afternoon passed a robin came and with a beady black eye sat over my hiding place and 'ticked' in alarm. The light rain stopped and a grey squirrel tip-toed along a branch above me and with tail a-twitch began to whine in a most agitated manner. And then at last I heard it – the sound for which I had been waiting – a series of snorting, grunting belches from a rutting fallow buck with a fine spread of antlers. He marked the area to be defended by fraying young trees and urinating under them. Then he dispersed the scent of this and of a special gland by means of the broad palmate antlers which gathered the scents during the fraying process. My buck began by issuing slow snorting challenges so that he could also listen to the calls of a young buck in the next territory. I could see half a dozen does standing under the beeches. As it became progressively darker his challenges became more rhythmic and sustained as he moved up and down the edge of the forest. An old grizzled dog fox trotted across a ride and I watched a newly emerged badger nosing his way along the fence. It was with great reluctance that I finally left this secluded spot to the belches of the fallow buck and to the pipistrelle and noctule bats that were hunting before dusk for their insect prey.

The heathlands of the New Forest reach to 419 feet along the Downton-Cadnam road and appear as parallel ridges; on Beaulieu Heath they form low plains. Reddish-purple bell-heather and mauve ling are the dominant plants of the heath. Common gorse gives a glowing golden appearance to the landscape in spring and there are also dwarf gorse and petty whin. Bracken grows everywhere and I have seen the erect spikes of spotted orchid and clusters of yellow cinquefoil and tormentil. Deer often graze the heaths at night with the domestic animals and return to the woodlands for the daylight hours. Hares and rabbits occur widely and there are field mice and voles. Silver-studded blue butterflies, small heaths and graylings flit over the heath in summer and there is the chance that one may come across lizards as well. I have found that the commonest heathland birds are linnets, tree and meadow pipits, wrens, wheatears with smaller numbers of stonechats and nightjars and in the boggy places curlew. There may be a few pairs of red-backed shrikes and woodlarks. The Dartford warbler – our only resident warbler, by the way – suffers in bad winters, and nearly all died in the severe winter of 1962–63 and it has also suffered from the destruction by fire of its very vulnerable habitat. 'Take Care. Do Not Start Fire' is the legend on the Forestry Commission notices, and apart from the destruction of trees and other living things the Dartford warbler is terribly at risk. The open woods and heath are also home for that dashing little falcon, the hobby, and a good year will see perhaps a couple of dozen pairs. The boggy parts of the heath are often full of bog myrtle and cotton-grass, bog asphodel, sundew and butterwort that trap the insects, occasionally gentians and the pale yellowish-green bog orchid, while Hampshire purslane now has its only British station in damp ground in the New Forest. In some of the wetter places and along streams grey wagtails, sedge warblers and reed buntings and a few pairs of snipe can be found nesting.

The New Forest heaths form part of a much larger area which sweeps in a narrowing corridor north of Bournemouth and Poole to Wareham and beyond to Thomas Hardy's 'Egdon Heath' country. These heaths may be warm and glowing with colour in late summer and yet damp and infinitely depressing in the mist and rain. Their flora and fauna may grade into those of the upland heather moors but these will be the subject of a later chapter. The survival of these southern heathlands depends very much on checking invasion by trees, but fire as a means of control, together with grazing by domestic animals, may lead to choking by bracken or purple moor-grass or the development of birch scrub. Many fires are accidental; an old heath may take twenty years or more to recover from the effects of deep burning. There are other hazards too, from reclamation with deep ploughing, clay extraction, atomic power stations and even on the Arne peninsula from oil. Where the heaths still survive, a wide landscape of brown wind-trimmed plain is lit up with purple for one brief period in the year, and dotted with gorse bushes, green-brown and only illuminated with gold in spring, solitary pines and some birch scrub, and pale green bog and marsh unfolds towards the horizon.

The Dorset heaths have plant communities similar to those in the maritime regions of western Europe reaching from France to Denmark. They are the home of bubbling curlews, singing skylarks, linnets, meadow pipits, stonechats, yellowhammers, wrens, green woodpeckers and woodlarks. There are also tawny and little owls, nightingales singing fit to burst and nightjars, 'lone on the fir branch . . . brooding o'er the gloom', to make the heathland nights as fascinating as the days. In Britain the rare and delightful Dartford warbler is almost totally confined to lowland heath, with its mature heather and well developed gorse bushes and even young conifer plantations. It is very interesting to the naturalist since it is a resident, whereas the other breeding species of warbler, apart from Cetti's, are primarily summer visitors. On warm spring mornings the little Dartford warbler, slate-grey and dull vinous in colour, will stand momentarily on the top of a gorse bush with his long tail cocked up for all to see before diving into the vegetation and re-appearing somewhere else. It is a great skulker but an enchanting bird. On wild and windy days it seems to disappear, completely swallowed up in the gorse. However, it is on the warm mornings that one is most likely to hear the song as the bird launches itself upwards in a little song-flight accompanied by a slightly metallic whitethroat-like warble, sometimes liquid, sometimes bubbling. I once spent part of a summer watching and tape-recording Dartford warblers on a Dorset heath, and in this region of heather and gorse there were also three pairs of stonechats, many meadow pipits and linnets, yellowhammers and curlews and a pair of grasshopper warblers. Both hobbies and Montagu's harriers hunted over the ground.

90

Certain parts of the heathlands – Hartland Moor with its rare Dorset heath and bog plants, and Morden Bog with dry open heath and cotton-grass, both not far from Wareham – are National Nature Reserves. At Arne to the east of Wareham there is a 680-acre reserve of the Royal Society for the Protection of Birds which has a small population of Dartford warblers. Harriers and merlins visit Arne, with its typical heathland birds, resident smooth snakes, slow-worms and four other native reptile species, interesting grasshoppers, dragonflies, spiders, scarce moths including the silver-Y which, however, came in thousands in 1973, and migrant red admiral and clouded yellow butterflies. The great drought year of 1976 set catastrophic fires alight on the southern heathlands. In Dorset more than one third of all the Dartford warbler territories were destroyed in one week in August including most of those on Hartland Moor. At Arne in 1978 there were only three surviving pairs. Fires on Surrey heathland in 1976 destroyed the habitats of many pairs of nightjar, woodlark and stonechat as well as some Dartford warbler territories.

These heathlands are perhaps the rarest and most rapidly decreasing of all the wildlife habitats in Britain and western Europe. Large tracts are in the charge of the military, which safeguards them from development, but the effects of tank and vehicle tracks and exploding missiles may to some extent offset the advantages. Fire and the effects of people and cars – 'access in excess' – may all change and degrade the heaths. Should the heathlands go on shrinking then their final survival may be, as Colin Bibby expressed it, as 'a few golf-course roughs, forest rides and quarry margins'. No naturalist doubts the need for the conservation of these heathlands but more money is required to ensure the survival of those swaths of heather and gorse that form a unique wilderness and a home for some of our rarest living things.

6. Concrete Desert

It may seem rather incongruous that in a book about the countryside of the British Isles there should also be a chapter on urban wildlife. Yet four-fifths of the population of England actually do live in towns and cities. Indeed for many of us the only 'countryside' that we regularly see are the playing fields and railway embankments along our commuter railway, the municipal park or urban churchyard in which we eat our lunchtime sandwiches, the grassy swards around our office or tower block, even our gardens. As a small boy I learned my natural history in London and still continue to observe and monitor the changes in suburban north-west London where I now live. I have never been at a loss in 'the Great Wen' to find something of wildlife interest even in seemingly unpromising places. My most recent revelation was that of a deep shaded groundfloor area at the back of a Victorian terrace in central London, completely surrounded on all four sides by high walls and windows; through a wide picture window in the firm's conference room I looked out on a small fig-tree and some climbing plants, where house sparrows were searching for insects and a blackbird had built a nest. Even in this world of concrete and brick, glass curtain-walls and tower-blocks there is still wildlife to be found. Of course, from the city centre outwards the number of species of wild plant, insect, bird and mammal increases. For example, where five may be the maximum number of different kinds of lichen on the gravestones in a city churchyard the number may rise to thirty or more in the suburbs and the lichens themselves act as indicators of the degree of pollution of the atmosphere especially by sulphur dioxide. Yet our tour of London should begin, I believe, in the heart of the city itself.

No longer do elephants roam the land where Trafalgar Square

now stands or hippopotamuses wallow and roar in the ancient tropical rivers or swamps. Instead the soils based on the London clay hold a smaller, less distinguished and invertebrate life – worms and mites. The city and its environs may attract various kinds of plant to their bits of wasteland and be the home too for many street pigeons – descendants of the rock doves once domesticated by man – as well as house sparrows, starlings, which nest in small numbers but roost at night in thousands on the buildings around Trafalgar Square or in the trees of the Parks, and kestrels. That attractive little falcon, the kestrel, can survive in the city on house sparrows and I have seen nests on the Savoy and Langham Hotels and on the Public Trustee's Office in Kingsway. Blackbirds too sing in Covent Garden and Lincoln's Inn Fields and even on Bush House in the Strand. In autumn migrants can be seen flying low over the city roofs – starlings, chaffinches, meadow pipits, lesser black-backed gulls and so on. Among the mammals there are brown and black rats, with the latter concentrated around the docks and Oxford Street but in a state of decline, house mice that can breed and live the whole of their lives in refrigerated stores, cockroaches and other insects, and spiders. I have seen a house mouse on the tracks at Piccadilly tube station, a brown rat at Holborn and a street pigeon on the platform at Marble Arch far below the ground.

There are many trees in the streets and parks and churchyards, especially London planes, which have a short leaf-season and shed their bark, giving them an advantage in conditions of soot and grime, horse chestnuts, sycamores, trees-of-heaven and limes which in high summer shower the city workers with their gyrating floating seeds. Alders have been planted near the Post Office tower in one of the pavements and seem to be doing well. Rosebay and Oxford ragwort are common flowering plants and both species were growing in 1978 on ledges on the National Gallery in Trafalgar Square, rooted perhaps in and nurtured by the droppings of street pigeons after seeds had been drifted there by the wind. When apples, cherries, figs and tomatoes appear on city wasteland wildernesses they are probably survivors of some office worker's lunch! One piece of wasteland that I watched became carpeted in its first year with rosebay, Oxford ragwort and groundsel and in the second year by bracken and coltsfoot. By the third year elder was established and twelve years later there was a considerable scrub growth of sallow, sycamore and ash rising to twenty feet or more in height. There were blackbirds, starlings, sparrows and pigeons as well as visiting tits, hedge sparrows and song thrushes.

In the Surrey docks some sixty acres of wasteland have arisen as the site became overgrown with weeds, bushes and small trees that flourished on the desert of brickdust and mortar. Here such 'countryside' birds as linnets, skylarks, yellow wagtails, sedge warblers, whitethroats, meadow pipits and even red-legged partridges have nested in the heart of London, and in 1970 two pairs of yellow wagtails – birds normally of marshy and watery habitats – bred on a bit of wasteland quite close to Vauxhall Bridge. Of course, the River Thames itself can provide a real attraction for naturalists with an occasional seal or dolphin a real bonus for the watcher, but there are often other delights – a great crested grebe at Westminster, even a razorbill, as well as Canada geese, starlings and meadow pipits and pied and grey wagtails on the river mud, mallard nesting on bridge piers and even some of the moored barges.

The Thames has always been more useful than picturesque. Nathaniel Hawthorne observed that the effect of the river was 'as if the heart of London had been cleft open for the mere purpose of showing how rotten and drearily mean it had become'. Now its banks are neater, with modern buildings, but some of these have taken away the effect of St Paul's Cathedral on the skyline and there are still eyesores to be seen. Its waters are cleaner so that it now boasts more than seventy kinds of fish, while salmon and even sea-horses have been recovered from the Thames. In fact, it is still one of the most important ecological factors in the natural history of London. From Hammersmith to Woolwich the banks of the river are largely built up and industrialized. Above Hammersmith the riverside is often lined with alders, willows and sedges and here from a boat I have seen redshank, Canada geese and even a scoter flying past. Below Woolwich there are muddy reaches like those of many east-coast estuaries. In one year more than thirteen different species of duck were seen between Fulham and Hammersmith, while at Woolwich several thousand pochard and hundreds of surface-feeding and diving ducks now take advantage in the winter of the cleaner waters of the Thames. To me the sight of geese flying above a dull grey river bearing its huge barges carrying London's refuse and made noisy by the passage overhead of helicopters and jet aircraft on their way to London airport is infinitely refreshing.

London stands on the Eocene clay which once bore broad-leaved woodlands. Today the city's forests are to be found in the Royal Parks. St James's is the oldest of these Parks. Henry VIII enclosed a swampy part of the Tyburn's water-meadows as a deer park and bowling green. James I introduced a collection of waterfowl and Charles II completely redesigned the Park. Now it is a small piece of countryside set like a green jewel in the heart of London. Here I find the geese and ducks just as pleasing as did John Evelyn, who called them in 1644 'a singular and diverting thing'. In winter the mallard quack and whistle, Canada geese approach from Buckingham Palace in a long descent to the lake, with loud honks that punctuate the roar of traffic in the Mall and Birdcage Walk. In May 1975 the Queen graciously allowed me to spend six hours birdwatching in the grounds of Buckingham Palace, and here among a glorious but controlled riot of azaleas and rhododendrons I saw 216 wild birds of no fewer than twenty-five species, including spotted flycatcher, bullfinch, chaffinch, jay, magpie, pied wagtail and the usual garden birds as well as nesting pochard. A pipistrelle bat has also been recorded by the Palace lake.

One December I stood by the bridge in St James's Park watching the birds and, whenever a visitor arrived to feed them, the black-headed gulls, perhaps from Poland, Germany or Finland, would rise up in a screaming flurry of wings, just as they used to do when as a boy I fed them in the 1920s. W. H. Hudson, who watched these gulls in St James's Park, described them as 'a picture of black winter and beautiful wild bird life'. For me it was always a special pleasure to stand by the lake among the tall plane trees, bare against the sky, with a wild chorus of geese, ducks and gulls in the background, even perhaps with the faint music of a distant military band in the Mall. I could also hear the rattle of magpies, the whistles of starlings, the deep 'chittucks' of redwings on the grass, the chirps of house sparrows clinging like dark fruit to the low branches of a willow as they waited for their daily benefactor, and, sweetest of all, the chiming note of a great tit – a promise of spring to come in London's countryside. In summer there are also the songs of blackbirds, the soft wheezes of blue tits, the chatter of starlings, the crooning of pigeons and even the long calls of the herring gulls nesting on the island to delight the walker in the park. In 1976 there were nineteen species of nesting bird and forty-one reported altogether. Hyde Park and Kensington Gardens, like Regent's Park, mustered thirty-five breeding species but these parks are better provided with shrubberies and sanctuaries. Butterflies are often to be seen, especially red admirals, tortoiseshells, small coppers and whites and even the painted lady.

Outside the city there lies a ring of inner suburbs formed from rows of nineteenth-century terraced houses, villas and mansions. This is often a land of grimy privets, laurels and aucubas where gardens are small and birds rather few – sparrows, starlings and perhaps a few house martins with occasional street pigeons and blackbirds. There are not many secure nesting places for birds while the often lopped and struggling trees may provide a nest-site only for a woodpigeon or a short-term roost for sparrows. In some of the older areas swifts sometimes find nesting places in the roofs of houses, gaining entry through a space left by a missing slate. A disused railway track runs through north London from Finsbury Park to Highgate, providing a linear kind of park some 1¾ miles long and twenty-six acres in size. I have plotted the plants and animals on the sometimes open, sometimes wooded embankment that winds its way through the rows of Victorian houses. Along the strip of brick-dust and rubble that once supported the metals are tall woods of oak, birch and sycamore and at least twenty-seven other kinds of tree. I discovered at least forty kinds of flowering plant and nearly thirty different species of bird, of which twenty were breeding. The nesting birds included mallard, collared dove, jay, mistle thrush, willow warbler, goldfinch, greenfinch, bullfinch and redpoll, many of which were birds of woodland origin. Foxes, hedgehogs and grey squirrels were living along the track as well as grasshoppers and many kinds of humblebee and other insects.

St Mary's cemetery at Kensal Green, with its ash trees and horse chestnuts, scrub and grassy swards, is another open space in Inner London which I have studied. Here among the gravestones with their daguerreotypes of the deceased and an 'Appian Way' of marble mausoleums there were two pairs of skylarks as well as brown, gatekeeper and blue butterflies – all within three miles of Marble Arch. In recent years some of the Victorian villas of the inner suburbs have been cleared and in their place tower blocks have raised their oblong ugliness above a once more gracious landscape. Swifts and house martins, gulls on their way to roost on the reservoirs and migrating chaffinches, larks and thrushes pass close to the serried rows of square windows. In summer, kestrels may nest on the window-ledges or in window-boxes providing eyries as high as the seventeenth floor! I watched a pair of kestrels and three young through a net curtain on the fourth floor of a block at West Ham. Other nests have been built on gasworks, power stations and tall chimneys.

Beyond the inner suburbs of London lies another concentric ring of buildings – middle class residential suburbs with semi-detached houses, red-tiled, pebble-dashed, sometimes gabled, with small gardens, built on land that was once farmland, market garden, orchard or golf-course. As the houses went up, so woodland birds like tawny owls, turtle doves, garden warblers, lesser whitethroats and tree sparrows moved out and so also did the pheasants, partridges, sparrowhawks, corncrakes, cuckoos, wrynecks and yellowhammers. The foxes and badgers, stoats and weasels went as well.

However, as the suburbs began to flourish and the trees and shrubs grew and matured so the gardens began to look like woodland and scrub once more, and there were for the birds the additional advantages of bare earth on the flowerbeds and short grass so that worms and invertebrates were easier to find. The new growth provided safer nesting, roosting and sheltering places and kind housewives fed the birds during the winter. Woodland birds did well – blackbirds, robins, hedgesparrows, tits, and in recent years jays and then magpies arrived as well, the latter now at plague levels and taking a toll of the first clutches and broods of the thrushes and blackbirds. Yet some of the species are now nesting in greater density in these suburbs than in their original forest home and the young of blackbirds are produced in greater numbers in London than in country areas.

I live on one of these inter-war estates and have studied the wildlife for nearly thirty years. The commonest bird is the house sparrow followed by blackbird and starling, while goldfinches, chaffinches, greenfinches, robins, song thrushes, great and blue tits, tree sparrows, hedgesparrows and woodpigeons have also nested in my garden. Summer at Dollis Hill is full of birdsong and the hum of insects while hedgehogs snuffle through the garden and bring their young to the backdoor. But other times of the year are equally fascinating. In early August I am often woken by the maniacal bursts of screaming that tell me that the common gulls have returned for the winter and are proclaiming the fact from a chimney stack. The bird-chorus of July has collapsed, and only the wren, diminutive but still full of passion, hurls his notes from the wild part that I maintain at the bottom of the garden.

Hive and humblebees come to my snowberries, Michaelmas daisies and clerodendrons, and occasionally a female leafcutter bee that chops out incredibly smooth oval sections from the leaves of flowers and trees. She is about two-fifths of an inch long with light brown hairs on the head and thorax, a pale fringe down the abdomen and slightly cloudy wings. On the underside of the abdomen is a light orange-red brush of hairs for collecting pollen. The cutting process is carried out with surgical precision. The bee flies in, grasps the edge of a leaf on one of my Frau Karl Druschki roses and audibly and speedily chops away at the leaf until she has removed an oval section. Once she sawed off the branch on which she sat, for as the piece of leaf came away she fell towards the ground and had laboriously to gain flying speed before she crashed to the earth. Each one of her brood cells is made up of oval sections of leaf. Some of the more interesting visitors among the insects to my garden have been painted lady butterflies, a silver-washed fritillary and a swallowtail

97

butterfly, silver-Y and swallowtail moths, red underwings, white plume moths and many lacewings as well as common field grasshopper and house cricket, emperor and common and brown Aeshna dragonflies and a damselfly that came to my plastic pond, to which I had introduced common frogs and to which the local common newts come quite unaided.

There is, as Shelley wrote, 'a harmony in autumn, a lustre in its sky' and there is quite a lot to observe at this time of the year. In the early morning a low sun glances between the boles of my fruit trees and casts a faint glowing haze over the grass now heavy with dew and dotted with fresh worm-casts. The silvery shining webs of orb spiders, built during the hours of darkness, shimmer as the breeze touches them with its delicate finger. There is a damp smell in the air and a robin's melancholy phrases, now gaining strength after August's break, are raised sweet and plaintive above the distant roar of the North Circular Road.

As the sun gains height, the grass begins to dry and the mowing can begin. After half an hour of labour I pause to rest and look up – and there is a small party of swallows winging their way purposefully south. The robin has long ceased to sing but a cock house sparrow on the roof – one of those that roosts at night in my philadelphus– provides a brief interlude of sound. A couple of moth-eaten blackbirds drop down to inspect the newly mown grass and a hoverfly hangs poised over a clerodendron flower; in 1976 a hummingbird hawkmoth was attracted by the purple umbels of flowers. As I finish trimming the edges of the lawn a hedgesparrow sings three phrases from a prunus before diving down to hunt for food at the bottom of the snowberry hedge. I start to cut the long grass around the foot of my fruit trees and the noise flushes out a party of starlings that had been feasting on my small pears; one flies up to the television aerial, readjusts his plumage and begins to sing. I start to cut the grass with shears and in a sudden flash of scarlet, white and black a red admiral butterfly takes off from a rotting apple and perches, resplendent, on a bramble spray. Wasps stagger up drunkenly from the fallen fruit or drift rather aimlessly around the garden as their life spans begin to run out. A migrant chiffchaff flits through the fence and from the tree above my head a blue tit stammers an uncertain phrase or two. I go on to tackle the weeding, with couch grass, groundsel, chickweed, swine's cress, willowherb, bindweed and rampant violets the chief enemies, but I remind myself that a weed is only a plant growing in the wrong place! Overhead a string of lesser black-backed gulls migrates towards the west and there are a few house martins flying and flycatching below them. A young hedgehog creeps out from under an ice-plant and onto the lawn, sniffing and moving forward with slow deliberate steps. He swallows a hard pebble of bread dropped by a careless sparrow. Then he starts to climb a six-foot high chainlink fence with ease, hand over hand; half way up he changes his mind and descends to make his way along the fence and disappear next door. I am relieved that Argus, my yellow Labrador, has not found him, otherwise he would return frustrated, bloody with spines in his nose and covered with highly active fleas!

Now tea is over and there is a chill in the air as the sun begins to go down. Parties of starlings are making their way to their roost in central London and a heron on broad slow wings flaps over to a night's fishing on the Brent Reservoir. The local blackbirds are coming to spend the night in the honeysuckle around the front door. The next-door cat, full of hope but without the skill to achieve success, makes yammering faces at them. It is the blackbirds' 'mik-mik' of alarm that I last hear as I leave the garden to the smoke of autumn and the coming night. Next morning I am up early to watch the suburban skies above my head, and into October too, since I often see large parties of starlings, skylarks, redwings, fieldfares and chaffinches flying north or north-west over the area after crossing from Europe on the same morning. They seem to prefer flying into a north-west wind. When the weather really deteriorates and Burns's 'flaky show'r Or whirling drift' begin their assault on London, redwings and fieldfares come into my garden to feed on the apples, and with them may be siskins, tree sparrows and even skylarks. Hundreds of larks, redwings, fieldfares, woodpigeons and lapwings and smaller numbers of linnets, chaffinches and sometimes snow buntings cross over my garden, fleeing from the snow.

In the twenty-seven years that I have lived at Dollis Hill I have observed no fewer than 115 different kinds of birds, including such rarities as the bittern, hobby, little ringed plover, greenshank, glaucous gull, hoopoe, greenish warbler and twite. Ninety-one of these different species I saw in or over my garden – an indication of the *rus in urbe* that I have found at Dollis Hill. The birds take advantage of the food that we provide, the cover that we allow to grow in our gardens and the rejectamenta of our society whereby plastic bags, wood shavings, cigarette ends, paper, string, burned felt and even metal wire can all be used in the construction of nests. Our dustbins, rubbish dumps and back gardens also enable many urban foxes to survive and even salute the moon from under some street lamp.

For many years one of my favourite haunts has been the Brent Reservoir, half a mile from my house and known locally from its former shape as the Welsh Harp. Built between 1833 and 1837 as a canal feeder by damming the River Brent, its natural banks and surrounding woods made it into a beauty spot, where such rare birds as night and squacco herons, little bitterns and spoonbills were recorded. Until the 1890s little change took place in the land use but by 1913 10 per cent of the area around it had been built on, and by 1970 the figure had risen to 65 per cent. Although diminished in importance it is still a refuge for wildlife and there are foxes, moles, shrews, squirrels and the Essex skipper butterfly. Reed-mace, reed-grass and reeds grow along the water's edge as well as flowering rush, various umbellifers and many garden escapes such as goldenrod, everlasting pea, chicory, Michaelmas daisy and snapdragons. At one end of the reservoir is a remarkable willow wood – a rare kind of habitat in Britain – which grew up after the water level of the reservoir dropped dramatically one year; in spring it is full of bird song, with willow warblers and sedge warblers making their contribution.

At the other end of the Brent Reservoir is a wood of oak and hawthorn – land for a cemetery that was never used but which was made into a local authority nature reserve – that brings the country-side into the heart of residential London and where I have listened to the songs in summer of blackcap, willow warbler, whitethroat and lesser whitethroat. I have walked through these urban jungles feeling that I was far away in some remote midland wood. On the water there are sometimes hundreds of mallard and pochard, tufted duck and coot, while I have also seen more than sixty gadwall as well as shoveler, teal and in winter smew from the Arctic. Great crested grebes – up to eight pairs now nest – and little grebes are regularly to be seen. There are many reservoirs, gravel pits and sewage farms in the London area, which represent the lost fenny, watery habitats that once were common in Britain.

There are many open spaces in the outer suburbs – municipal parks, allotments, factory surrounds, scrub, rubbish tips, ceme-teries, railway sidings, playing-fields and so on. Each may have its own small population of interesting plants and animals. Allotments have the advantages for birds of cultivated earth set amongst strips of grass and weeds rather like the squares on a chessboard. At the top of Dollis Hill is a layer of glacial drift left at the time of the maximum glaciations which provided a fine friable soil; shortly a deep con-creted reservoir will be built here and then light-weight housing put on top. By 1974 only one fortieth of the acreage given over to allotments in 1961 had survived; this was due to the policy of municipal in-filling that reduces the open space in areas where it is perhaps most valuable in our urban context. Scrub will often grow around industrial sites and along railway embankments, and I am always advocating the planting of berry-bearing shrubs to delight us and help to feed the birds.

One railway bank near my home supports sycamores, regenerating elms, birches, robinias, broom and bladder senna, while in their respective seasons everlasting pea, yellow toadflax, yarrow, mugwort, comfrey, hoary cress, horseradish, coltsfoot, cypress spurge and Aaron's rod, rosebay and Oxford ragwort, flower and add beauty to the iron way. One such bank, left uncut by London Transport at my request, has colonies of field grasshoppers, common and holly blue butterflies, small skippers, commas, small coppers and meadow browns – all countryside insects that in London bring me pleasure and reward. Rubbish tips are a world of their own, enticing gulls, crows, starlings, sparrows and street pigeons, house crickets and wasps. The gulls, which have increased enormously in recent years, forage on the dumps and roost safely at night on London's reservoirs, which are important too for ducks and wading birds on migration. Wetlands are decreasing all over Europe and the lakes, natural or man-made, ponds, gravelpits, and sewage farms of the London area help to provide alternatives for some of our most attractive birds. Pipistrelle bats can often be seen hunting above the reservoirs and pits along the Thames Valley.

So far I have traced the wildlife of London from the city nucleus out through the concentric rings of development to the very edge of the real countryside. Such a system of division is quite impossible for, say, Birmingham with its adjoining built-up zones, where various townships have been fused together, forming a 'jumbled mosaic' of development. W. G. Teagle in *The Endless Village* has described how the commercial centre of Birmingham lies quite close to the villas and woods of Edgbaston, and how the Black Country has 'a remarkable array of wild woods, rocky precipices, marshy pools and grazing land enmeshed amongst its towns, expanded villages, factories and furnaces'. Our busy overcrowded towns are not always the deserts that they first seem. There is little doubt too that the variety of wildlife is an indication of how we treat our environment. If there is a diversity of living things then they reflect an environment in which we ourselves can flourish and our more spiritual needs be satisfied. The wild places in our towns and cities are deserving of close examination and possible conservation. In their own way they are almost as significant and precious as some of our better known reserves – the green acres of the New Forest, for example, or that classic piece of wetland at Minsmere in Suffolk.

7. Reeds and Reflections

It was a still cool morning as I walked between the tall brown waving reeds, damp and somewhat disconsolate in the autumn greyness. From secret hidden meres deep in the reedbed came the staccato honks of coot and the deep quack of mallard. A hedgesparrow in a dwarf sallow piped in alarm and a snipe rose calling 'scape-scape' as it went off, surprising a couple of water pipits. A sound of squealing broke out from the reeds as a water rail began to call or 'sharm', making noises like a frightened piglet. The track now curved around a more open pool and I paused to gaze across the water with its silvery reflections of reeds and sky. A kingfisher was sitting on a bending reed stem peering into the water, and a flight of mallard on their long approach broke away at the last moment as they saw me on the track below. Three mute swans were dibbling and muttering together and I could hear the pinging, ringing notes of bearded tits, tawny brown to match the reeds. Suddenly the brown shape of a male marsh harrier came beating over the pool; all the birds fell silent for a moment and then the chiming notes of the bearded tits announced their departure. A moorhen gave a warning 'curruc'. Here I was in a world apart – a fragment of our history almost untouched by the twentieth century.

Once fens occupied vast tracts of Cambridgeshire, Huntingdon, southern Lincolnshire and the area of the Humber. Although the name Fenland still survives for that area of land lying south of the Wash, all else is changed. In February the now fertile but eroding acres of the Fens lie bleakly under a leaden pall of clouds pregnant with rain or snow or bathed in sunlight under a sky of the clearest blue. Neat reed-fringed dykes and drains drive their straight lines across the black peaty ploughlands with their rich yields of sugar beet, potatoes, celery, carrots and corn. Only from the air is it possible to trace the paths of the old meandering rivers. Once this was a fenny wilderness of trackless swamps and meres and a waste of peat bogs and islands. This was the home of the monster Grendel and 'Black Shuck', the spectral descendant of the Black Hound of Odin. Alders took root in the clumps of tussock sedge, reached maturity and died. The weight of trees in some places made the peat sink and then, as the roots encountered deeper water, the alders could no longer survive. Underneath lay a quaking floor of tussocks and black stinking pools. Where the floor was more stable, downy birch, ash and oak might appear amongst a jungle of wild raspberry, currants, bramble, dewberry, privet, buckthorn, guelder rose and a field layer of hop, meadowsweet, nettle, yellow flag, marsh marigold and purple loosestrife; the fen water draining off the chalk accounts for the lime-loving plants.

The Romans seem to have tried to drain the Fens and Tacitus noted that 'the Britons complained that the Romans wore out and consumed their bodies and hands in clearing the woods and embanking the Fens'. The Isle of Ely really was an island and Coveney 'the island in the bay'. Hereward the Wake reached the sanctuary of Ely along a causeway, and in 1695 Celia Fiennes reported that it was still the only way into the city. This was a land of malaria and rheumatism where the sovereign remedies were thought to be alcohol, opium pill and poppy tea. In 1490 John Morton, Bishop of Ely, built a straight cut or leam to stop the River Nene flooding the Isle but it was not until 1630, when the Earl of Bedford was given a grant by Charles I, that serious drainage under the direction of Cornelius Vermuyden could begin. For the next two centuries the drainage proceeded, often in the shadow of disturbances and riots among the fenmen or 'slodgers', 'camels' because they crossed the swamps on stilts or 'yellow-bellies' because they lived like frogs. One of their complaints has survived:

Come brethren of the water and let us all assemble
To treat upon this matter which makes us quake and tremble,
For we shall rue it, if't be true, the Fens be undertaken
And where we feed in Fen and Reed they'll feed both beef and bacon.

Such were the fears of the fenmen who, according to the historian William Camden, 'apply their minds to grazing, fishing and fowling'. However, with the draining of Whittlesea Mere in 1850 the long process of drainage came to an end and now there are only a few tiny maintained remnants such as Wicken Fen – a sedge fen owned by the National Trust with a remarkable community of rare plants and animals, and Wood Walton Fen, a National Nature Reserve where I have seen fen violet and elecampane growing. There are also the Broads, which have been called 'the greatest reservoir of wetland flora and fauna in Britain', Minsmere in Suffolk and a few small fragments which can give us some idea of what Fenland was really like.

North-east towards Denver lies a great memorial to the Dutch engineer Vermuyden – the Old and the New Bedford Rivers. Between them are twenty miles of wash or flood plain to take the overspill of water, the outer banks being raised above the level of the inner ones. The washes are used as safety valves where floodwater can pile up and relieve pressure on the surrounding fenlands. When the pressure of water gets very high, sluices are opened which allow some of the water to run into the washlands. In summer cattle, sheep and horses graze the region between the rivers, known as the Ouse Washes, enriching the land and so making it more attractive to the wildfowl that come in fantastic varieties and numbers in the winter. Large areas are looked after by the Cambridge and Isle of Ely Naturalists' Trust, the Wildfowl Trust and the Royal Society for the Protection of Birds. In summer black-tailed godwits and ruffs – regained nesting waders – come to breed on the Washes. Black terns and little gulls have nested as well and there are countless skylarks, reed buntings, sedge warblers, yellow wagtails and rarer species like garganey, gadwell, pintail and shoveler.

Fenland in early winter is often dull and forbidding, with a leaden misty pall lying over the reed-fringed dykes, the black peaty ploughlands dotted with lapwings and the last sparrow-haunted stubbles where tractor engines give a constant reminder of the loading and carting of sugar beet. On one such day the bleakness of the black lands between Ramsey and Chatteris was underlined by the sad wailing screams of common gulls – sounds with a truly primeval quality. Fenland has a unique appeal for me, with the villages balanced on their gravel knolls, the yellow brick houses, the white herring-boned church steeples, the farmsteads standing in their little oases of trees, the herons motionless in the dykes, the mallard and wigeon flying over high on rapid wings, the weasel racing across the road in front of the car and the collared doves singing in the townships and around the grain-dryers. Away in the cold, hazy distance the tall grey towers and lantern of Ely Cathedral beckon across the Fens. Hundreds of fieldfares flash grey rumps as they move their feeding grounds over the farmlands. I once made a special journey to the south of Ely to see the Old Engine at Stretham built in 1831, whose steam worked a scoopwheel that drained the Waterbeach Level and is now preserved for all to see.

Vermuyden's drainage scheme was at first a great success but then the peat dried out and began to shrink so that the level of the land dropped below that in the drainage channels. Water had to be pumped up to the level of the rivers and cuts and in the eighteenth century several hundred windmills, used to do the pumping, dotted the Fenland landscape. Later, steam, diesel and electric pumps were introduced. At Mullicourt Priory, as a result of the fen drainage and the shrinkage of the peat, the water courses are at different levels and here, on the Middle Level Main Drain which shows the difference in level very clearly, I saw a Continental black-bellied dipper searching for food on the water's edge. Flocks of rooks and jackdaws were feeding in the fields when a sudden blast of Arctic air caught them unprepared and sent them scurrying into the sky, tumbling and falling like black leaves, in their efforts to achieve mastery over this surprise blow, their voices clamouring in the wind.

As winter in fenland grows colder and damper the wildfowl begin to flock in. Bewick's swans often pass over, honking as they go. I travelled along a narrow fenland road where black, tanned bog-oaks, dragged from the nearby fen, were stacked along the verges, silent witnesses to the past floods which had brought about their doom. I came at last to a muddy track which rose up the bank of the New Bedford River. Cautiously I peered over the top and a water shrew swam away into the cold waters of the river. In front of me were hundreds of acres of washland and waterlogged meadow and shallow pools, criss-crossed with drainage ditches and dotted with shrubs. There were at least four hundred Bewick's swans from northern Russia, yellow-billed and dibbling in the water. Some were muttering musically and conversationally between themselves, some were waving wings and displaying and others were landing or taking off. I could also see great numbers of coot, moorhens, mallard, wigeon and teal with several shoveler and handsome drake pintails. Towards me from the swampy, ready wilderness came the musical bubbling 'hoos' and 'hohs' and the more staccato 'kaa-kays' of the Bewick's swans, the harsh quack of mallard, the slurred whistles of drake wigeon and the growls of the ducks, the musical 'krit-krits' of teal and the scurry of coots, the plaintive cries of lapwing and curlew – a picture in sound of one of the great wildfowl refuges in the British Isles.

Although the Fens are still famous in places for their wildfowl there were losses when the original watery wastes were drained. As the regions of willow and alder-fringed reedbed and mere were cleared, so also went the greylag geese, perhaps the original breeding stock of the Suffolk goose trade. At its peak in the 1780s as many as ten thousand geese could be seen being driven through Chelmsford on their way to market in London. Other losses were the black-tailed godwit, ruff, bittern, black tern, spoonbill, Savi's warbler, the native large copper butterfly, the rosy marsh moth and most of the swallowtail butterflies. The tall and handsome crane bred until about 1600. With a growing awareness of the need for conservation in Britain and essential protection, godwits and ruffs, black terns, bitterns and Savi's warbler have returned to breed on the Ouse Washes or elsewhere, and in 1967 the rosy marsh moth was rediscovered but this time in North Wales.

The black peat fens of today were formed from freshwater floods from inland rivers when sedges, reeds and other water-plants died to form the peat. Closer to the Wash the low-lying land was once regularly flooded by inrushes of the sea, which left a mixture behind of sand and gravel. Above the sandy fields of these drained silt fens the skylarks sing in summer and white butterflies and furry humblebees buzz and drone. Here there are orchards and beds of roses. May is the month of the tulip – that exotic flower from the Court of Suleiman the Magnificent, Sultan of Turkey, which probably crossed the English Channel about the year 1572. The tulip fields are very attractive and have some interesting wildlife.

I often begin my tour at Wykeham Abbey, the private chapel of a fourteenth-century prior of Spalding. Its gaunt shell stands on a low mound where I have watched cross-bred Charolais cattle grazing in a framework of vivid strips of flowering tulips – golden-yellow, cream, lemon, scarlet, brick-red and magenta. Cheerful women, bent double, pick off the flowerheads, leaving rows of pea-green stalks and leaves. Their act of decapitation gives strength to the bulbs and an almost unlimited supply of flowers to decorate the floats in Spalding's Flower Parade. Yellow wagtails, brilliant in the sun, balance on the surviving scarlet flower cups and house and tree sparrows chirp from the oaks and horse chestnuts along the field's edge. A narrow reed-fringed watercourse separates field from field and the Abbey from the road; here baby moorhens, tiny balls of black fluff, call anxiously and are answered by sharp 'kittucks' from their parents. Pheasants call from the banks where red-legged partridges scuttle and hide, and brown hares lope idly over the arid sandy ground.

I am glad to see that horses can still find a place in the bulbfields especially when the ground becomes heavy after rain. Men gather cauliflowers from a rich jungle of nettles, chickweed and veronica while a single corn bunting sings from a taut wire stretched above the field. A few yards away the River Welland flows brown and sluggish

106

while common terns, white in the sun, hawk up and down the water in a flush of spring excitement. West of Spalding lies Vernatt's Drain – a channel named after Sir Philibert Vernatti, a Dutchman who helped to drain Deeping Fen between 1632 and 1637. Mallard and shelduck drift slowly on this drain and sand martins fly back and forth giving their dry stony chatters. At the time of Spalding's Flower Parade some eight million tulip heads, each fastened to a straw-covered metal framework, are pinned on to the features of each float, perhaps a swan, dog, leaping horse, camel, snail or ladybird. One float I saw bore nearly 200,000 tulip heads alone. The parade slowly winds its way on a circular route around the town in early May – floats, ancient cycles, girl pipers, attractive girl models, bands and all. I have watched the parade from outside Springfields, the bulb industry's magnificent spring garden that lies north-east of the town. After one parade I walked in these gardens and listened to and tape-recorded a lesser whitethroat rattling away in song and quite oblivious of the public address system, the traffic and the hundreds of visitors surging past his temporary song post!

Minsmere on the Suffolk coast is really a man-made fen, having been flooded in 1940 as a defence against Hitler's tanks. It is now a vital sanctuary for many fenland birds and one of the most famous and important reserves of the Royal Society for the Protection of Birds; my old friend the late James Fisher called it 'this glorious Suffolk marsh'. Here behind the seawall are many acres of reedbeds, marsh, mudflat and gravel 'scrape', woodland and field; in all it claims a total of more than 110 kinds of breeding bird with nearly a hundred nesting in a recent year. Each year about 210 species of bird are recorded and the all-time record is around 280. I have made many visits to Minsmere throughout the year but some in early April have been the most rewarding. Each spring morning I would be out walking on the marsh among the waving reed fronds although these were not as tall as some I have negotiated in the Camargue or the Coto Doñana. On the edge of a favourite clearing with alders and willows on one side, sedges and open water on two sides and dense reedbeds on the fourth, wrens and hedgesparrows sing against a background chorus of crowing pheasant and shy booming bittern in the reeds. Marsh tits and the first male chiffchaffs are proclaiming their swampy territories. A male marsh harrier – and Minsmere is the last toe-hold in Britain for this fine, long-winged bird of prey – mounts up into the sky to a height of a thousand feet or more. From here it launches into a series of spectacular dives, climbs, plunges, loops and sweeps but in a constant pattern of rises and falls. This rare spring display is accompanied by a characteristically plaintive cry. A mole with shiny spring coat emerges to spend a couple of hours digging shallow tunnels in the waterlogged ground and a stoat runs across a fallen tree trunk lying across a dark ditch. One April morning I watched two spoonbills – one adult, one juvenile – come in a broad sweep over the reedbeds to pitch in a shallow pool; here they stayed to feed until the resident mute swans in their ire at seeing other white birds drove them off in a flurry of mud and water. As the sun warms the air on April days a chorus of deep rolling croaks and short plaintive barks, like canine punctuation marks, arises from the pools and dykes; spring has come to the frogs and toads and to Minsmere itself.

In May grasshopper warblers and garden warblers sing in that ecotone which lies between reedbed and scrub, and it is worthwhile listening for Savi's warbler and even Cetti's warbler. Hundreds of pairs of reed warblers sing in the reedbeds and their continuous summer chatter is perhaps the commonest sound. Water rails squeal and garganey crackle in the forest of reeds while in the waving carpet of green there is a growing population of bearded tits. Bitterns boom and sometimes walk out of the reeds before slipping back into their special wilderness, or clamber laboriously up the bending reed stems or flap slow and brown over the pool to seek new foraging grounds. Marsh harriers quarter the reedbeds ready to pounce on some small unsuspecting mammal, bird or frog. The reeds stir gently in the summer wind and from a tangled mass of scrubby plants a sedge warbler sings, in Ralph Hodgson's words, 'His old, old ballad new'. On a swaying reed plume a black-headed, white-collared reed bunting sings his metallic refrain – 'Jink-jink-tillee'. Invisible among the reeds on a secret tiny pool, beloved by the coots, are several little grebes whose high whinnying trills and bubbling laughs form duets that rise and fall on the warm summer air. Cuckoos call in the distance and pheasants crow and flap their wings. On the lagoons and purpose-built gravel islands terns and avocets nest and shelduck indulge in their displays with soft masculine whistles and loud laughing feminine cadences. I have watched ospreys hovering and side-slipping after fish in the Island Mere, black terns and one year a white-winged black tern with white rump, tail and shoulders.

In September I have spent many hours hidden in a simple reed hide and later a wooden one watching the muddy flats, which lie flanked on all sides by reeds, where the migrant waders come – redshank with fluting calls, spotted redshanks calling 'Tchuit' or giving subdued trills, greenshank with low triple whistles, whimbrel, curlew, little stints, curlew sandpipers, godwits, ruffs, green and wood sandpipers and ringed plovers. From my low reed hide I could see little stints and curlew sandpipers walking within a few feet, snipe probing with long bills, a green sandpiper running about with a wood sandpiper for comparison and a single heron trying one day to swallow a large eel and on another a flounder nine inches long and about five wide! Near the sluice by the seawall one September I watched more than sixty blue tits, a dozen willow warbler/ chiffchaffs, six blackcaps, two whitethroats and a rare barred warbler. When winter comes to Minsmere the mammals that inhabit the marsh – stoats, weasels, rats, mice, voles and occasional otter – go about their affairs in comparative secrecy, unlike the coypus. The pools and meres are visited by noisy honking Canada geese and snow-white Bewick's swans from Siberia. Teal dibble in the shallows, wigeon are busy flighting, pochard and tufted duck drift like paper boats on the Island Mere, black-headed gulls scream incessantly, water rails still 'sharm' and bearded tits chime and kingfishers whistle as they fly in to land on a reed stem, bending it down in an elegant curve towards the water. Coot and moorhen busy themselves in the reedy border of the mere, a heron calls 'fraank' and I can see a marsh harrier and, beyond, above the fields, a hen harrier, all grey and black. As the big red sun begins to sink towards Eastbridge, the starlings arrive to roost in the reeds. Soon the evening air starts to pulsate with the calls and murmuration of thousands of birds as they begin to perform their massed aerial manoeuvres – one of the most dramatic of all the sights in the British countryside in autumn.

I have talked at some length about the Ouse Washes and Minsmere because to a very considerable extent they recreate the old fenlands, but there are, I am glad to say, other watery wildernesses but not as many as conservationists would like. There are the Norfolk Broads – the flooded sites of early medieval peat workings which are linked today by river channels both of which are often fringed with bog and marsh – a former stronghold of wetland flora and fauna in Britain. With the development of shacks and shanty towns, the growth of boating and disturbance and the pollution of the water it is rather the confiding common bird species that have survived at the cost of the rarer. The deoxygenation of the Broadland water giving to it an unhealthy milky cloudiness has adversely affected the fish and other aquatic life and the birds as well, and its present condition gives rise to serious concern. Some of the broads have well-known names: Hickling – a National Nature Reserve where the Norfolk Naturalists' Trust has provided birdwatching facilities; Horsey Mere – more brackish, where I recorded the strange moans of the coypu; Hoveton Great Broad – a by-passed broad with its white and yellow water-lilies; Oulton with its water sports; Wroxham – empty of birds in the summer but dotted with wildfowl in winter.

There is a tremendous variety of watery landscape in the Broads, ranging from busy recognized waterways to quiet meres and marshes. Lakes, sometimes open, sometimes wooded, with alder carrs and ashwoods or dense beds of reeds, ornamented with yellow iris, water violet and water soldier, are linked together by winding rivers. Yet with a quarter of a million visitors each year the pressures are causing Broadland to die. One perhaps thinks of autumn in terms of birch trees etched in fragile gold and beeches in burnished glowing copper, but for me one of the most glorious riots of colour is not among forest trees but in a secret jungle of Broadland scrub. Several times in autumn I have been to one of the few private broads in a region of drowned peat cuttings, swamps, meandering water-ways, dense scrub and, where the land rises, belts of oak, ash and hazel. Dr E. A. Ellis – Ted to his friends – has rowed or punted me along watery channels with a reedy margin of loosestrife, meadow-sweet and iris all long past their blossoming. Buckthorns flourish along the bank with a few leaves turning to gold and a harvest of blackberries. Yet the overwhelming beauty of this place is in the banks of guelder rose with their leaves a-glowing, flaming red and their shiny red berries hanging in glossy bunches like fat redcurrants, well loved by redwings, fieldfares and waxwings. I could see a few spindletrees and there were marsh, willow, great and blue tits foraging in the alders and willows and some diminutive goldcrests in an oak. Siskins and redpolls visited the alders, bullfinches piped in the sallows and far away I heard some mallard and a water rail. A skylark was still singing overhead and, as I looked up to listen to him, I saw two cormorants flapping dark and menac-ing across the sky. Swallows and martins passed by on migration and the first redwings of the winter were exploring the scrub with a few song thrushes. Robin song was everywhere and I heard jays and both great and lesser spotted woodpeckers calling close by. A heron got up in front of the boat. There were many signs of coypus – those large aquatic mammals that escaped from East Anglian fur farms and were living wild in the Broads, frightening yachtsmen with their strange calls. After trying to re-establish the extinct large copper butterfly Dr Ellis found that whitethroats were eating the bodies of the females! At Wood Walton Fen large copper larvae have been reared on great water dock in muslin-covered cages.

In summer the Broadland alder carrs have all the common species of tit, wrens, robins, blackbirds, song thrushes, chaffinches, reed buntings and pheasants, three kinds of woodpecker, treecreepers, tree sparrow, redpoll, bullfinch and summer visiting chiffchaffs, willow warblers, blackcaps, cuckoos and turtle doves. Tawny owls and jays are frequent while crows and one or two pairs of long-eared owls breed in the more remote carrs. In many springs there are reports too of the golden oriole.

I have visited other reedbeds in Britain and Ireland, and there are some fine ones around Lough Neagh, as we saw in Chapter 3. Bitterns boom and bearded tits call at Walberswick in Suffolk and near Cley on the North Norfolk coast; in this latter region I have seen on the reedy pools ruffs, green and wood sandpipers, pectoral sandpipers, a single Terek sandpiper and even a red-necked phalarope. The Royal Society for the Protection of Birds leases or owns Strumpshaw Fen in Norfolk – a habitat for wetland birds and rare insects, including the swallowtail butterfly whose preferred food plant, milk parsley, is very common. Great crested grebes, herons, kingfishers, mallard, teal, coot and moorhen can all be seen on the broad and nearby river, while sedge, reed and grasshopper warblers nest together with bearded tits, reed buntings and water rails. The fen carrs at Strumpshaw hold woodpeckers, nuthatches, woodcock and many birds of woodland habitat. Coypus are common and have to be controlled, and there are occasional otters to be seen. Some of the more unusual plants include marsh pea, marsh fern, water soldier and marsh sow-thistle. There is also Leighton Moss close to Morecambe Bay which has the only regular pairs of bitterns breeding in the north of England as well as reed warblers, water rails, teal, shoveler, pochard, tufted duck and sometimes garganey. The four-spotted libellula and brown aeshna dragonflies are common and large numbers of damselflies occur as well. I have been to the Moss but I shall defer a fuller description of it to Chapter 10.

Reeds often occur by reservoirs, gravel pits, lakes and on the landward side of coastal lagoons and river mouths. They are sometimes planted to improve the habitat; this happened at Sevenoaks Gravel Pit Reserve – the brainchild of the late Dr Jeffery Harrison and his father – and here in Kent more than thirty different species of wildfowl have been recorded as well as such rare birds as aquatic warbler and purple heron. At Middle Fen in Suffolk bog bean and marsh cinquefoil grow with marsh helleborine orchids, and at Titchwell in Norfolk there are reed warblers and a dozen pairs of bearded tits in summer and hen harriers and short-eared owls in winter. In the diminishing reedbeds at Fairburn Ings near Ferrybridge in Yorkshire more than a million swallows and sand martins used to roost in the autumn.

For me the tall waving reeds transport me at once to a special

111

world of delight and satisfaction. There are the sibilant hiss of reed stems and plumes bending and returning in the wind and the strange smell of damp and rotting sedge vegetation. Such a world is the complete antithesis of our modern urban style of living. In the summery chatter of warblers and the boom of bitterns and in the wild voices of swans, ducks and waders in spring and autumn there is a reminder of a lost primeval world. This impression is only confirmed by those biting, stinging insects of which both Samuel Pepys and John Evelyn complained. Reedbeds and meres, scrub and carrs provide one of Britain's rarest and most important wildlife habitats and each small piece of fenland, wild and reedy, is a treasure for us all to guard and enjoy.

8. Vistas of Shingle and Sand

It has been said that in any civilized country the two kinds of habitat least changed by man are those of the mountains and the coast, and I believe that the seashore of the British Isles with its variety of flora and fauna can have a very special appeal and interest for the naturalist. The sea, of course, has a profound effect on our shores, rolling rock fragments into rounded boulders, pebbles or sand, cutting away at cliffs and shifting material; in broad terms it can be said that any particular coastline is one formed by erosion or deposition. The work of the sea is very much achieved through the action of waves, with huge masses of water being moved and charged with large amounts of rock, stones, gravel and sand. On a shelving beach a pebble is thrown up parallel to the direction of the prevailing waves but is dragged down once more by gravity roughly parallel to the slope of the beach. So pebbles work their way along the shore to continue their progress or to be impeded by lines of man-made groynes.

Sea-cliffs speak of erosion but shingle beaches and spits, often adorned by sand dunes or encompassing mud-flats and saltmarshes, tell a story of deposition. Indeed, more land in Britain is being gained from the sea than is being lost by erosion. As the pebbles are moved by the action of the waves along the coastline they often form spits away from the original shore but at right angles to the direction of the waves; good examples are Hurst Castle spit in Hampshire while at Dungeness there are successive lines of shingle beaches. Behind the shingle spits muddy saltmarshes often form. On sandy shores wind plays an important part, bodily shifting the sand, and dunes may grow in association with the vegetation. New dunes are yellow or white in colour while grey dunes are older and, as soil has begun to form, usually have a closer plant cover. Original hollows in the dune scenery may be aggravated by wind action and sometimes small depressions are made larger by blow-outs. There is a constant change, a living dynamism, about the complexes of shingle ridges, dunes, sandbanks and mudflats which occur around our shores.

Perhaps the finest example of this multi-faceted scenery is to be found in North Norfolk. I have stayed in this part of the world many times over the last quarter of a century and each year I have found a different picture as dunes blow away, storms ravage the beaches and floods burst in from the sea. I have experienced the soft days of spring, the intense heat of summer, the gales of autumn and the bitter blizzards of winter, learning to know, love and understand this remarkable part of the world and its people. There is the phenomenon of Blakeney Point, joined to the land at its eastern end near Salthouse with a shingle spit formed by the movement of beach material to the west; in conditions of westerly winds there is some slight irregular shift of stones in the opposite direction. The ridge itself gives rise to a number of recurved ends which allow a narrow entrance by the tide to a broad area of marsh inside. Into these shallow marshes the tide flows and stands for a while before the ebb, depositing a great deal of mud which, in turn, is attractive to many wildfowl and wading birds. The Marrams, the Hood and the Headland ridges are all recurves brought about as wind and waves turned the spit landwards. Sand dunes grow on the shingle ridges especially at the Headland. Through the kindness of the National Trust I have been able to film for television and tape-record the Point and its wildlife on a number of occasions. One May morning the warden, Ted Eales, took me across in his boat from Morston on the high tide. As I landed on the shingle ridge in blazing sunshine, reflected from the parabolic curves of the nearby dunes, skylarks were soaring overhead in their song flights and meadow pipits were parachuting down in their displays. Ringed plover ran away over the stones, common and little terns screamed over my head and rushed about carrying tiny fish in their bills and I spotted four black terns speeding purposefully across the water.

Here was an exciting world to explore. On the strand line I could see plants of sea rocket and prickly saltwort, sea purslane and oraches while higher up the shingle there were clumps of shrubby Suaeda and marram. Many kinds of grass had taken a hold and I was delighted to find some sea holly and sea wormwood. The dunes themselves showed development from an inch-high hillock lodged against a blade of grass to great rolling banks of sand. Sea couch, marram and sand sedge carpeted the dunes and among them I found sand forget-me-nots and whitlow grass. The saltmarsh between the Point and the land revealed a succession of vegetation from the sea rushes and red fescue of the higher regions down to belts of sea pink, sea purslane and sea lavender which can turn the marsh into a blaze of colour, then down again to the acres of Spartina rice-grass, sea blite, perennial glasswort, sea aster and Salicornia.

One evening in May I took my recorder to the Far Point to wait for the incoming tide. I had a fine view over a huge expanse of mud and sand on which I could see a motley collection of wading birds and a single Brent goose, all feeding in complete silence. There were whimbrel, godwits, redshank, turnstone and dunlin as well as ten knot with two of these waders in their glorious chestnut breeding plumage. As the tide came within thirty feet of the shingle spit three oystercatchers arrived, their piping trills gradually accelerating and changing in pitch and volume as the seconds went by. Small parties of graceful and attractive common terns – veritable sea swallows – started to beat up and down over the shingle ridge and incoming tide in buoyant and confident flight. Their long grating screams dominated the scene. Half a mile away a colony of noisy black-headed gulls was building in a wide patch of Spartina that clothed a muddy arm pointing out into Blakeney Sound.

June at Blakeney Point is very much the month of the terns. One year my wife and I visited the Point in the second week of the month. It was a morning of hot yellow sun as we concealed ourselves behind one of the sand dunes. Above us myriads of wings showed up ethereally white against the deep blue of the sky. Sandwich terns with black crests and shallow forked tails were coming back to the colony with silver sand eels for their downy grey and spotted young. In the clear air their black crowns and yellow-tipped black bills made a magnificent sight, while the dunes echoed and re-echoed with their strident 'quirricks'. Many adults were standing guard over their young or muttering and stabbing at chicks which had wandered out of their own territory into another. There were clutches of common terns' eggs lying in unlined shallow scrapes in the sand; here the adults chattered and shrieked, displaying their long tail streamers and black-tipped scarlet bills.

Behind the dunes lay the tidal waters where the waders flock and, when the sea is flowing in, the rarer little terns appear – miniature 'fairy' terns with yellow bills and white foreheads which fish above the incoming tide, hovering with rapidly beating wings and head fixed still before plunging down into the water for a sand eel or shrimp or breaking off at the last moment as the prey swims deeper.

115

When their foraging is done they assemble on a sandy spit to rest, preen or call with sharp 'kik-kiks' and softer 'kirri-kikkis'. Redshanks had built their nests in tiny clumps of marram grass. Oystercatchers were sitting on eggs in the shingle ridges and ringed plovers had laid their clutches in tiny depressions among the stones. Almost wherever we carefully picked our way ringed plovers would get up calling 'tu-lee'. There were many shelduck and we saw one pair on the water bravely trying to marshal fifteen ducklings which stretched in a line some forty yards long like a battle squadron. Sometimes we could also hear the whistle of a displaying drake and the answering laugh from his mate. From our vantage point in the dunes we could see grasshoppers and earwigs, a few spiders and a number of butterflies including wall, grayling, small heath and small white.

I cannot exactly recall how many separate Septembers I made my base at Cley near the eastern end of the Sound. A naturalist should walk, as I have done, the three and a half miles along the shingle ridge from the Cley beach road to the Far Point. As you walk over the shingle the pebbles chatter, whine and shift tiringly underfoot but I have been rewarded by strange mirages in summer and the sight of many rare birds in winter, spring and autumn. The village of Cley-next-the-Sea – all flint with a large well-preserved windmill and the great church of St Margaret's, with its Decorated nave and brasses around whose lighted windows Daubenton's bats regularly hunt – has been one of the focal points of my life as a naturalist for many years. My base was the George Hotel where Mrs Eve Burnett – a charming and attractive hostess – ran a real birdwatcher's home from home. The shingle ridge at Blakeney contributed to the decay of Cley, Blakeney and Wiveton as ports; a record of 1582 shows that these three ports then possessed seven ships of more than one hundred tons whereas Yarmouth only had four and King's Lynn two! Now Cley is a mile from the sea.

In September the coastal marshes have become a sanctuary for large numbers of wading birds each bringing its own special music to North Norfolk – the whistles of curlew, the trilling notes of turnstones, the high-pitched 'ker-reep' call of flying dunlin and the flutings and lesser notes of whimbrels, godwits, ruffs, redshank, spotted redshank, greenshank, curlew sandpiper, little stints, ringed, golden and grey plover, while the water itself drains musically away from the evergrowing area of sand and mud. Here the cold wind races across the mudflats and only the bent bait-digger or the hardy birdwatcher braves the elements. Not far from Cley is a small remote marsh where greenshank find security, and I have seen both green and wood sandpipers on its muddy edge. The saltmarshes are bright with the rayed flowers of sea aster. In the belts of Suaeda that lie behind the shingle banks I have seen bluethroats, wheatears, stonechats, whitethroats, icterine and barred warblers, ortolan and little buntings and red-breasted flycatchers. Above the coast passing Arctic skuas chase the Sandwich terns, forcing them to disgorge their last meal, and offshore gannets, red-throated divers, grebes and the first wigeon are passing. On the arable fields within a stone's throw of the shingle large numbers of linnets, greenfinches, bramblings and reed buntings search for food with a handful of Lapland buntings newly arrived from the far north.

By October there are snow buntings as well along the shingle ridge and one or two hooded crows while a hen harrier is often busily quartering the fields above the marsh. Skylarks still go on singing and there is a steady passage of birds along the coast – divers, kittiwakes and terns going east and scoters, eiders and wigeon travelling west. Sometimes huge North Sea breakers come pounding up the shingle bank hurried by a wind that picks up the snow buntings in its blast and scatters them far and wide like snowflakes in a gale. One morning I came across a large dog otter searching for crabs and shellfish at least two hundred yards out on the mudflat. Above him a screaming pack of herring gulls danced up and down until in his own good time he made for the cover of some Suaeda bushes. That same day the wind turned to the south-east and very low over the North Sea came a stream of migrant redwings and song thrushes to drop exhausted into the first clumps of marram that they could find to rest; they were so tired it was possible to pick them up by hand. With them came some robins, redstarts and a bluethroat – just one of the many migratory movements that can be seen in autumn on the North Norfolk coast.

Winter here can be almost unbearably severe. When conditions are bad rock pipits come into Cley village and bitterns in the freshwater marsh near Salthouse can be seen feeding out in the open. I once arrived on the Blakeney ridge during an Arctic gale after crossing in an open boat, to spend several nights in a small wooden hut on the Hood. The sea was freezing and thousands of knots, weaving about like smoke in the sky, pitched finally on the firm mud to join other wading birds trying to survive in the bitter weather. Brent geese and shelduck were dibbling among the frost crystals. The wind strengthened and twice in the darkness it was necessary to move the boat up the ridge. A giant tide and a phenomenally fierce north-west gale assaulted the shingle ridge on both sides, finally punching a hole in the ridge itself. On the voyage back to Cley in a raging blizzard and

117

with all colour taken out of the landscape I passed three different kinds of diver, mergansers, scaup, goldeneye, velvet and common scoters, smew, tufted duck, white-fronted and pink-footed geese, Brent geese, Bewick's swans and a single glaucous gull all seeking shelter in Blakeney Sound. The Bewick's swans, beating their way into the teeth of the gale, swept overland and then out in a wide arc to disappear in a flurry of snow far out beyond the savage breakers on the shingle ridge. But it is not always like this in winter. I have stayed in February in the old lifeboat house at Blakeney Point, on that first Nature Reserve in Norfolk which was handed over to the National Trust in 1912. I have walked outside in the early morning to listen to the songs of skylarks and meadow pipits overhead and the notes of linnets in the dwarf trees near the Point, and to see thirty seals hauled up on the sand bar. I have hidden myself in a watery clump of Spartina to await the arrival of the Brent geese from the sea in order to record their resonant 'krronk-krronks' before being rescued as the tide began to lap round my place of concealment. This coastline is a fascinating one, subject to great physical change but also one of the finest complexes of sand and mud flat, shingle, dune and marsh to be found in this country.

To the west of Blakeney lies Scolt Head Island, much of it also purchased by the National Trust. It arose as an offshore bar, extending to the west by beach drifting, and forming a series of recurved ends. There are many large dune systems covering parts of the main ridge of shingle, which itself appears to form a section of a complex of other ridges at Brancaster and elsewhere. The succession of plants is rather similar to that at Blakeney. The mottled and the lesser marsh grasshopper both flourish at Scolt and many of the moths are basically coastal insects with their larvae feeding on marram, sea couch grass, thrift or sea pink and sea lavender. There are many kinds of snail and marine animal, and more than a score of different kinds of beetle feed on decaying plants and animals on the tide line. There are rabbits, rats and field voles, and Scolt Head Island is also famous for its breeding terns. To many people the seashore in mid winter may seem a rather desolate, even forbidding place. I have always been a beachcomber and my home literally bursts with the bric-à-brac and rejectamenta of both man and nature – pebbles of every shape, colour and texture, bones and prehistoric artifacts, sea-smoothed fossils, shells, luscious deep green glass floats, bottles encrusted with acorn barnacles, claws and carapaces of crustaceans, wings and feathers – trophies from the shores of the British Isles.

One of my favourite winter expeditions is across the broad sands of Brancaster Bay, opposite Scolt Head Island. The beach is clean and firm to walk upon and during the autumn high tides the salmon swim silver over the golden strand. At low tide pebbles and shells are sprinkled over the sand, each isolated and easy to see against the background of the yellow beach. Shallow runnels of water meander across the shore and, as the water popples and drains away, rows of black-headed gulls stamp up and down on the strand, curlew pass high overhead out of gunshot range and parties of dunlin and ringed plover swing past – small flashes of grey and white. Scattered across the sands are the flat wrinkled plates of oysters, shading from snow-white to grey brown. There are the shells of horse mussels up to five inches in length; some are black and gold while others flake like an old oil painting and show flecks of Tyrian purple. Razorshells lie broken with their hinges gone. There are huge whelks, common mussels, variegated scallops, blunt gapers, fluted cockles and the pink and green fragments of crab shells. In a few places, spreading up the golden beach, are fan-shaped sooty stains from the drifted scraps of countless black mussel shells. Where the mussel debris piles up, oystercatchers pipe and trill and suddenly there is a blink of white and grey-brown wings and a party of snow buntings rises from the tideline of sea wrack, sponges, mermaids' purses, cabbage leaves, dead gulls, plastic detergent bottles and rusty tins. From far out across the flats come the husky barks of the greater black-backed gulls. As I wander further along the beach I come across occasional black lumps of wood or compacted peat, washed up from the ancient submerged forest nearby. Some are as big as footballs and many are riddled with the neat holes of various boring creatures such as gribbles or shipworms.

Near Brancaster Harbour there are patches of shingle composed of rounded or fractured chert and flints, white quartzite pebbles, pieces of soft red sandstone, lumps of brick and concrete and ovoid lumps of chalk, tunnelled by small marine animals and easily scratched with a finger-nail. I have found pebbles of glass like abandoned sapphires, emeralds and topazes, no longer transparent, and frosted with tiny pits and chips over their surface. I watch out for the dull red pebbles of jasper – an opaque kind of quartz – and of citrine and carnelian. Around me as the diligent search goes on, the redshanks flute and bob and by Cockle Bight there are several hundred Brent geese. As you come closer to the geese their heads go up and then they mount up into the air with a chorus of rolling calls – a true sound of the North Norfolk coast in winter.

I have only seen Gibraltar Point in Lincolnshire and across the Wash from the air when I have flown low over the great sand banks looking for pink-footed geese. At the Point there are marram-covered

119

dunes and sandy areas where sand sedge has taken hold. Here hound's tongue, stonecrop and sea buckthorn flourish and the adjoining salt and freshwater marshes have a flora that ranges from sea aster to marsh horsetail and pyramidal and early purple orchids. There are many kinds of butterfly and a wide range of birds on this local Nature Reserve and regional Wildfowl Refuge. It is also an important migration point, with a Bird Observatory. Terns, shelduck, ringed plover and redshank breed. To the south there is the great shingle ridge and foreland of Orford Ness, where beach material is moved southwards along the Suffolk coast. At Shingle Street there are vast ridges of pebbles all roughly parallel, and in the words of Suffolk's own poet, George Crabbe,

> Where all beside is pebbly length of shore,
> As far as eye can reach, it can discern no more.

There are considerable patches of vegetation and one of the most characteristic plants of this shingle spit, where I have sometimes seen wheatears nesting, is the purply-blue sea pea which flourishes among sea campion, and yellow horned poppy. Scrub and grassland have colonized the landward side of Orford Beach. The keep of Orford Castle, built by Henry II to guard the Haven and constructed from London clay, coralline crag and Caen stone, looks out across the diverted River Alde and Havergate Island, managed by the RSPB as a reserve for wildfowl and waders. Orfordness and Havergate form a National Nature Reserve. Havergate was once part of Gedgrave Marshes and was probably formed by continuous stages of silting up and breaking open of fresh channels by the Rivers Alde and Butley; as far as can be judged its banks were built in the first half of the fifteenth century. These man-made defences to the island were breached by the great storm of 31st January 1953 but the banks were repaired and the island drained in time for the return of the avocets – Havergate's pride and joy after their absence from Britain as breeding birds for more than a century, and reappearance in 1947. I have travelled by boat down Butley Creek in the early morning to see the avocets but there were other breeding species as well – Sandwich terns, redshank, oystercatchers and ringed plover. Black-tailed godwits, dunlin, knot, greenshanks, ruffs, wood and curlew sandpipers and little stints can be seen at migration time, and in winter there are many species of duck – shelduck, wigeon, mallard, teal, shoveler and pintail. Kestrels are often present on Havergate and in 1977 took many of the avocet chicks which survived the first few days of life; that year, with its cold weather and high winds, no avocet young were reared on the island.

At Dungeness in Kent shingle forms a great triangle of land that pushes its way out into the English Channel. There are other spreads west of Rye and again at Hythe, but it is the mass of shingle at Dungeness that literally takes the breath away. It is formed from a large number of individual ridges each of which was once the outer, seaward edge of the expanding promontory, built up of shingle piling up from the west. The area – some ten thousand acres – carries a variety of vegetation, including grassland, which is used for grazing the famous Romney Marsh sheep. The Marsh is a wild region of dykes, reeds and meadows. In the *Ingoldsby Legends* 'the world according to the best geographers is divided into Europe, Asia, Africa, America and Romney Marsh'.

I have frequently made my way laboriously over the desert of rounded pebbles that move and chatter under one's boots around the point of Dungeness itself. Large areas are quite bare while in other places there are communities of flattened broom and blackthorn, lichens, and low clumps of elder, gorse, bramble and hawthorn. The skyline is dominated by the great grey-white slabs of the two atomic power stations which destroyed the physiography of this national asset and unique piece of European shingle. The old lighthouse, from which I carried out migration watches and which was built in 1904, has been replaced by a slender pencil of modern design and fully automated workings. Not far away there are miniature mountains and ranges of shingle piled up by grading machinery, draglines and vacuum plants as man withdraws from Dungeness the materials to satisfy his insatiable demand for building and concrete. There are new pools where the draglines grab huge mechanical handfuls of shingle from under the water but leave or shape new islands for the birds to rest on. The warden of the RSPB works with the contractors, and the Bird Observatory is assisted with finance from the Central Electricity Generating Board. Long conveyor belts clatter away into the distance bearing their harvest of stones.

There are some interesting breeding birds including wheatears that nest in old ammunition boxes sunk in the shingle, red-legged and common partridges, lapwings, ringed plover, hedgesparrows, skylarks, terns and perhaps stone curlews. On some of the flooded pools there are breeding coot, moorhen, mallard, grebe, sedge and reed warblers. In autumn I have seen parties of common and black terns on the shingle islands with greylag geese and grey plover, the latter sometimes still in Arctic plumage. Linnets sing fragmented snatches of song from the gorse, goldfinches dance over the thistle-heads and there are often migrants in the brambles – whitethroats,

121

willow warblers, redstarts, whinchats and thrushes. Many of the blackbirds, song thrushes, redwings and ring ouzels passing through Dungeness pause to feed on the blackberries which are remarkable for their size and delicate flavour. I have seen hundreds of goldcrests in the gorse bushes and once I watched three of these tiny birds sitting on the tops of old oil drums. In the cabbages at the foot of the old lighthouse in Mr Dowsett's garden I counted nine black redstarts and a reed warbler. I have, of course, enjoyed several nights of bird migration from the gallery of the old light when, illuminated by the rotating searing beams, redwings, song thrushes, starlings, robins, lapwings, golden plover, wheatears, chiffchaffs, goldcrests and firecrests were drawn towards the light. The air around the lighthouse was full of sound and these were some of the most exciting and satisfying nights of my life. Now the old light functions no more and the migrants are not attracted in the same way and, although the column of the original lighthouse was lit by floodlights to avoid casualties, some birds used to be killed.

The sky at Dungeness that looks down on the shingle, the power stations, the lights and the huts is immense at any time of the year. The end of the foreland is a fine spot to watch passing ships and migrating terns, kittiwakes, Arctic skuas and visiting little gulls. One summer day I could see dolphins off the Ness busily hunting mackerel that were, in turn, feasting on shoals of whitebait that left a silvery strand along the shore at least a yard wide. I have watched starlings coming in to roost in the willow scrub that grows on the damp shingle, but of all the nights that I have enjoyed at Dungeness I remember especially one when skeins of high cirro-stratus cloud flared up into strands of bright vermilion and purple, soon to grow more vivid and a more brilliant crimson near the horizon. I could hear the sea on one side while a common sandpiper called from a pool, half a dozen curlew whistled and two whimbrel gave their seven calls, a redpoll trilled overhead and two bats began to hunt over the shingle ridges. I stood remote and over-awed by the sunset in that strange stony desert. The wild call of a greenshank, down from the flows of the far north, set the final seal on my day in the strange autumn landscape of Dungeness.

In some ways equally fascinating and intriguing is Chesil Beach – a long unbroken ridge of shingle running from Bridport to the Isle of Portland eighteen miles away. Beyond Abbotsbury it leaves the coast and curves smoothly south-west, embracing a lagoon called The Fleet between itself and the coast. This strip of water varies from two hundred to a thousand yards across. The pebbles of the beach are graded in an extraordinary manner, increasing in size from north-west to south-east but apparently decreasing *below* the water in the opposite direction. Its origins remain something of a mystery; beach drifting does take place but the directions are confused. Perhaps the Chesil Beach originally formed a spit. Professor J. A. Steers wrote that 'in the light of known changes at Orford Ness and other shingle formations, the likelihood of a chequered past history for Chesil is more than probable'.

Chesil Beach is the first stretch of coastal shingle that I came to know. It is a high bank with steep sides standing as high as thirty feet or more above the water level of the Fleet. There is little vegetation but as the water in shingle beaches is generally fresh and plentiful plants do not suffer from drought conditions. In 1911 during a very dry period sheep came down to feed on the sea campion along the Beach. Little grows on the lower part of the seaward-facing ridge and although many halophytes, which can live in soils where the water is salt, grow on shingle, strong winds and salt spray may inhibit the establishment of seedlings. Storm ridges occur on the seaward side and vegetation is confined to the stable crest and landward side of the beach. Shrubby seablite – *Suaeda fruticosa* – is common and grows along the drift line of the Fleet behind the beach and along the ravines or 'cans', along which shingle is moved down by sea water forced through the beach. There are also sea campion, sea beet and orache, purple cranesbill and planted tamarisk, and I have also recorded sea and sharp rush, teasel and Dorset heath. Ringed plovers and oystercatchers breed along the beach and in past years I have seen common and little terns, while there are also reports of Arctic terns in recent years. There is a colonial nesting site for mute swans at Abbotsbury and this is well worth a visit in the breeding season. It has existed since at least 1393, having been created by the monastery whose tithe barn still exists as proof of its existence. Some artificial control of the swan numbers has taken place in recent years.

There are other notable areas of shingle and sand in Britain. Ridges of shingle on a smaller scale can be found at Langley Point or the Crumbles near Eastbourne, at Slapton, which we visited in Chapter 2, near Porthleven in Cornwall, Shoreham Harbour, Appledore, in Orkney, Shetland and the Isles of Scilly. There are also some fine sequences of sand and shingle in Cardigan Bay and on the southern coast of the Lleyn Peninsula. The dune system at Morfa Dyffryn on the Merioneth shore arises from a sandy foreland that has grown towards the north. It is a National Nature Reserve; a region of wind-blown sand with marram, red fescue and wild thyme, while in the damp hollows there are acres of creeping willow with varying amounts of horsetail, sedge, marsh pennywort, sharp rush and marsh orchids. No breeding birds are present in the mobile dunes but wrens and reed buntings visit them in the winter months. In the dune hollows or 'slacks' there are three regular breeding species of bird – skylark, meadow pipit and lapwing. In winter mallard, snipe, curlew and redshank can be seen on the 'slacks'. The northern end of Morfa Dyffryn has been described as 'possibly the best dune land-scape in England and Wales' and certainly one finds a strange mixture of wildness and remoteness. Not far away is Morfa Harlech which reveals new dune formations and is a rather complex region which arose by the formation of marsh and sand-flats within a spit.

On Anglesey, Newborough Warren – a National Nature Reserve since 1955 – is a magnificent stretch of sand dunes and marsh with panoramic views of Snowdonia. There are mobile marram dunes, where shelduck, curlew, skylark and reed bunting nest and ravens, wrens and redwings and fieldfares come in winter, and 'slacks' where white bent is the pioneer colonizer followed by a dense growth of creeping willow. There are also colonies of terns and strong passages of waders – redshank, greenshank, ruffs, godwits, little stints and curlew sandpipers. Farther south in Wales the shingle and sand formation of Ro Wen nearly blocks the estuary of the River Mawddach. Dawlish Warren has a double sand spit across the mouth of the River Exe and here in spring the sand crocus raises its pale purple petals, and the sea rocket blossoms along the strandline. Inspired by the writings of the late Henry Williamson I visited Braunton Burrows in Devon a number of years ago. Saltwort grows on the high water mark and the seaward-facing dunes are covered in sea couch grass. There is a large dune system and some of the highest marram-covered dunes are eroding away but the 'slacks' are carpeted with white bent, sharp rush and creeping willow. On the more stable dunes moss is common as well as such plants as dovesfoot cranesbill, storksbill and rest-harrow. This too is a National Nature Reserve provided with special trails.

There are also sand dunes in Aberdeenshire and there is also the machair – the shell sands of the Western Isles – which appear in Chapter 16. Further shingle and sand formations occur on the Moray Firth with dunes, sandy flats and marshes. Perhaps the best known of the Moray complexes are the Culbin Sands and Bar. Much of the area is now State Forest with Corsican pines growing and fixing the sand which once slowly engulfed the countryside. Today the newer dunes carry marram grass, the more settled sandy areas sand sedge and heath rush, and the fixed dunes ling and heather. Birdsfoot trefoil, mouse-ear, heath bedstraw and ragwort flourish on the stable zones. The marshes are sandier than those of North

Norfolk and support sea meadow grass, sea pink, sea milkwort, sea aster and marsh arrowgrass. On the newer marshes glasswort and annual seablite appear while prickly saltwort and sea rocket grow along the drift line. Findhorn Bay nearby is a great stopping off place for wildfowl and waders and I have watched flocks of mallard, wigeon, teal, with smaller numbers of mergansers, while the salt marsh which has developed between the four-mile long Culbin sand bar and the shore is again a refuge for mallard, wigeon, common scoter, teal and less regularly shelduck and velvet scoter. I have seen foxes and roe deer in the Culbin area and there are many small birds including meadow pipits, wheatears, mistle thrushes, coal and crested tits near the coastal conifer plantations while skylarks, little terns, Arctic terns, common terns, sparrowhawks and buzzards can be seen along the dunes and shore – a varied list of birds.

For me and other naturalists the features of the shingle and sand spits and their adjacent marshes are unique in the British Isles. There are the great vaulting width of the sky, enormous landscapes of grey, brown and gold, silver and green, the sound of the sea and the calls of waders and wildfowl in one's ears. Each region of sand and shingle has a different setting and this is an intrinsic part of the attraction. In Norfolk the old coastline with its trees, grey church towers and flinty red-roofed houses frames the shingle ridge and salt marshes. At Scolt Head Island there is a flat unimpeded view across Missel Marsh to Brancaster Staithe with its yachts and flapping stays. Mudflats are full of intricate patterns of winding channels and straighter waterways with pale fringes of sea purslane; the mud is often curved and convoluted in sinuous grey folds or glinting blue or green in the changing light. Sometimes the marsh is pink as in May and June, blue-grey later with sea lavender and sea aster. Serried ranks of conifers provide a deep green backcloth for the dunes at Culbin, Newborough Warren or Morfa Harlech, while Dungeness lies open but dominated by man-made structures and noises, the products of twentieth century technology. On the other hand Morfa Harlech looks across to the Castle built by Edward the First and standing on its stony mound of Cambrian rock. My feeling for the remarkable scenery that sand and shingle offer which has attracted to me places as far apart as the Isles of Scilly and the Isle of Lewis arises from an appreciation of all the varying facets and features of the landscape, seen in their relationships of time and space and as an intrinsic part of all that is enriching and satisfying in the countryside of the British Isles – an amalgam of sights, sounds and understanding that the naturalist is privileged to experience and to enjoy.

125

9. Cambrian Hills

Wales is, of course, distinct from England, cut off by the Marches and the courses of the Rivers Dee and Severn. For the naturalist it is a beautiful and fascinating country, with its mountain massifs, unfolding moors, streams and lakes, high woodlands and remote valleys. The naturalist R. M. Lockley said that the form of Wales, characterized by her geography and geology, 'was born out of earth and fathered by weather'. The geology of Wales is a complex one in which igneous and sedimentary rocks have been folded and denuded while the subsequent landscape has been shaped by ice action which plucked out the corries or cwms of the mountains from Brecon north to Snowdonia and left behind often enchanting lakes or llyns. A single chapter in a book cannot possibly do justice to the variety of countryside and fauna in a region as large as Wales, but since hills and mountains are often of the greatest interest to naturalists I propose to travel the Principality along the higher ground from the Rhondda north to Snowdonia and the Lleyn Peninsula.

There is much of interest in South Wales. There is the Gower Peninsula of Glamorgan with its Carboniferous limestone cliffs and sandy and rocky bays, its Stone Age cultures and the famous Paviland cave with its prehistoric animal remains. Although its limestone is a playground for the people of Swansea it boasts the unique yellow whitlow grass as well as vernal squills, spring cinquefoil, goldilocks and hoary rock rose among its flowers, while there are breeding gulls, auks, fulmars, rock pipits and wheatears to delight the birdwatcher. Balancing the Gower on the north coast of Wales is, of course the Carboniferous limestone in Anglesey, Caernarvonshire and especially at the Great Orme with a ridge running to Denbigh and Ruthin. In fact, I know the Great Orme's Head better than the

126

Gower and have travelled the toll motor road to look at the cliffs, the quarries and the plateau. Here a spring succession of flowers unfolds – vernal squill, spring cinquefoil, hairy violet, hoary rock rose, rock pepperwort, sea cabbage, and later still, sea pink and sea campion, samphire, bloody cranesbill and many kinds of hawkweed. On the top of the Head are wind-trimmed shrubs of wild cotoneaster, privet, hawthorn, juniper and common buckthorn. I also look out for the rare spiked speedwell and dark red helleborine. Kestrels hunt over the Great Orme, herring gulls wail, and a chorus of groans and caws comes up in summer from the auks and fulmars breeding on the great cliffs. Butterflies are common and the little cistus forester moth is only to be found in Wales, on the rock roses growing on the Great Orme.

But it is time now to return to South Wales – to 'little England beyond Wales' – in fact; this is Pembrokeshire with its Coast National Park, its castles of Manorbier, Carew and Pembroke itself and its romantic and fascinating islands. These include the RSPB reserves of Ramsey with over thirty regular nesting species of bird, and Grassholm, which holds fifteen thousand pairs of gannets and, viewed from the sea, looks in the breeding season like an iced sponge cake. I have also been to Skokholm – the site of the first British bird observatory – with its thirty-five thousand pairs of Manx shearwaters and its nesting storm petrels, and to the island of Skomer which is a National Nature Reserve. Again on Skomer there are some ninety-five thousand pairs of shearwaters as well as other seabirds, grey seals and its own special race of bank vole; I have walked amongst drifts of bluebells, red and white campion, vernal squills and primroses while the songs of wrens, larks, sedge warblers, blackbirds, wheatears and pipits have risen above the cacophony of kittiwakes, gulls, guillemots, razorbills and puffins on the cliffs. The mew of buzzards and the rough notes of choughs make Skomer for me one of the most enchanted islands off the coast of Britain. Not far away is Dale Fort – the mainland Field Centre of the Field Studies Council, which undertakes wide research and teaching projects and arranges trips to some of the islands. The cliff walks run from St David's to Strumble Head – a turning point for many night cross-countries that I flew during the last war – and there are also the Presely Mountains which form smooth grasslands and a home for buzzards, ravens, curlews and wheatears; from the igneous Ordovician rhyolites and dolerites in this hilly group came the bluestones for Stonehenge. Yet my survey of the Cambrian Hills had better be confined to the mountainous backbone of the Principality which runs due north and south.

127

In South Wales within a rim of hard sandstones and conglomerates of the Old Red Sandstone and Carboniferous limestone lies the basin of the famous coalfields. Inside the resistant Millstone Grit are the fields of coal formed from the trees and ferns of the Carboniferous Period. A thick layer of sandstone known as Pennant grit separates the famous secluded narrow valleys such as that of the Rhondda and provides wild stretches of moorland which rise in places to a height of two thousand feet. For me there always seemed something rather incongruous about the tall smoking chimneys and pithead wheels nestling in the folds of the blue-grey moors. To the north of Merthyr Tydfil, beyond a belt of limestone, the Old Red Sandstone gives to the Brecon Beacons National Park its warm, russet colouring. In the parish of Vaynor there are sheltered woodlands of oak, ash, wych elm and birch along the river. The yew and ivy-covered cliffs of the Ddarreu Fawr give way to moorland fringes; there are sheets of fresh water at the head of the valleys which often attract mallard, tufted duck, pochard, and goosander and I have sometimes seen whooper swans as well. The moorlands and peat bogs of Cilsanws and Pen Moel Allt are home for wary red grouse, snipe, curlew, wheatear, meadow pipit, skylark and occasional ring ouzel and even merlin, while in March I have seen short-eared owls hunting the moors. My old friend Geraint M. Lewis who lived in Vaynor and knows this region well wrote in 1955 that 'we can take pride in our Buzzards, Falcons and Ravens with their superb mastery of the air high above the hills which so prominently dominate our rural and picturesque parish'.

The Brecon Beacons are mountains, unspoilt and largely uninhabited. Within the confines of the National Park there are not only wooded regions of oak but also the extreme westerly beechwood in Cwm Clydach and the limestone outcrop of Craig y Ciliau, with its intricate system of caves and its beech, lime and unique whitebeams. The Old Red Sandstone crags of Craig Cwm Du and Craig Cerrig Gleisiad represent the most southerly limit of some arctic-alpine plants in Britain and there are moorland and peatbog plants as well. The moors are clothed in sheep's fescue grass, bents, moor and mat-grass, crowberry and bilberry, heather and ling, heath bedstraw, tormentil and heath milkwort, while the peatbogs are white with cotton grass that blows in the mountain wind above the reddish sticky rosettes of the flycatching sundews. In summer the streams have sandpipers, grey wagtails and dippers and among the grazing sheep wheatears disport themselves and red grouse call 'go-bak'. Ring ouzels pipe plaintively and it is always possible to spot a raven, buzzard or occasional merlin searching for food over the moor. On

128

the crags at Craig Cerrig Gleisiad kestrels, stock doves, ravens, ring ouzels – and rather oddly spotted flycatchers – nest, while the breeding birds of the moors and grassland also include cuckoo, stonechat, whinchat, wheatear, wren, meadow pipit and pied wagtail and red grouse. Between 1970 and 1977 non-breeding visitors to this National Nature Reserve included teal, sparrowhawk, kingfisher, wood warbler and crossbill, while among the other birds seen were peregrine, raven, fieldfare, redwing, black redstart and Lapland bunting. This is a land of sun and shifting cloud – a rich and beckoning region for the field naturalist, the hill walker and the lover of open spaces. I have also wandered through a number of the small fragments of wood in the dingles and crevices of the Welsh hills and moors; at the lower levels there are pied flycatchers, redstarts, tits, thrushes, willow warblers, tree pipits, great spotted woodpeckers and buzzards and sparrowhawks to greet you, while ravens, crows and willow warblers can be found right up to the highest level of growth.

To the east of the Brecon Beacons beyond the River Usk lies the range of the eastern Black Mountains, and to the west the Carmarthen Van and Black Mountain whose towering red sandstone cliffs partially enclose Llyn y Fan Fach – a lake which when I viewed it from the air looked like the water-filled shaft of an ancient volcano but is, in fact, of glacial origin. This beautiful llyn has drawn to itself a legend about a fairy maiden who rows a golden boat with silver oars on the surface of the lake. After marrying a farmer who promised not to strike her three times with iron she bears him three sons but with growing affluence from her fairy cattle he breaks his promise and she then disappears with her stock into the lake. Later she returns to teach her three sons how to heal people and leaves behind a legend not only of herself but of a local race of physicians. On the striated cliffs and fans of scree around the llyn grow mossy and purple saxifrage, roseroot, globeflower and rare ferns, while water lobelia and sweet-smelling bog asphodel flourish in the water and damp places. Due east by a couple of miles is Llyn y Fan Fawr, another glacial lake behind a moraine, and in its environs are buzzards, merlins, ravens, red grouse, dippers, grey wagtails and common sandpipers and, if one is very fortunate, a passing peregrine, perhaps a young male driven on by wanderlust.

From Brecon I have driven quite often to Builth Wells, through Lower Chapel with its noisy rookery and along roads made golden in spring with glowing patches of coltsfoot. Another road runs from Brecon to Llandovery, parallel with the old Roman road. Ten miles to the north where the Gwenffrwd and Towy valleys meet there is a region of superb mountain scenery; it is now a reserve of the RSPB and covers an area of more than a thousand acres. Along the roads and rivers are small fields, some with hedges; these are displaced on the slopes by bracken-covered sidings. In the steep ravines are hanging oakwoods with ferns and mosses where one can listen to the songs of pied flycatchers, redstarts and wood warblers, the mewing of buzzards and the deep croaks of ravens, while along the streams there are the sandpipers, dippers and grey wagtails that we met in the Brecon Beacons. Ring ouzels sing their wild notes or call sharply on the moors, and where heather flourishes red grouse stake out their territories. There are badgers and foxes, red squirrels and polecats. These small Mustelids with black and white faces and small white-tipped ears are dark above and creamy below. They live in open country with scrub woodland, bogs and rocks and may well out-number the stoat in Central Wales. Polecats have been increasing in south-west Wales, where they have been recorded well into Pembrokeshire and Carmarthen as well as the Usk Valley and the borders of Monmouthshire and Brecknockshire.

For the naturalist this region is of especial importance since it is here that he or she may see over the hanging oakwoods a buzzard-like bird with long narrow wings and a deeply forked tail with white patches under each wing, soaring and tilting its tail from side to side. This is the red kite in its last breeding stronghold in the British Isles, where perhaps eighty birds still hold on. In Oxfordshire in the 1820s according to O. V. Aplin 'the kite was abundant' and Shakespeare earlier still had underlined the widespread distribution of the bird when he wrote: 'when the kite builds look to lesser linen', since he was aware of its habit of taking rags, paper and even washing off the line to build its nest. The kite's main food is carrion, which is why it was so familiar to medieval Londoners; this it searches for over the ffridd – the marginal hill-land and the mountain sheep-walks; but the bird has suffered from the attentions of egg-collectors, disturbance, the afforestation of its old haunts and poison put out for foxes. Red kites desert their nests very easily and investigation of suspected breeding must be avoided at all costs. This was one pressing reason why I have never wished to make tape-recordings of this pride of Wales when other sound recordings of Continental red kites existed.

The kites will forage over quite a wide area and I have seen one over Afon Claerwen, which leads me into my next stop in Wales – the Elan Valley. I first travelled in this part of the world early in April when the countryside seemed to be sleeping in the unshakeable torpor of winter and only the daffodils in the villages below had given a hint of spring. A misty rain was falling over the reservoirs and gathering grounds from which the city of Birmingham gets it water supply. Very few people live in this part of Wales but over the moors graze some forty-five thousand sheep owned by Birmingham Corporation and hired out to tenant farmers, who pool their efforts at shearing time. The Caban Reservoir has sinuous curves and is bordered by grazing land and conifer plantations. Pen-y-Garreg Dam is dominated by a central cupola which looks down on a bubbling cascade of white water, while the Claerwen Dam towers into the sky – a mighty memorial to the engineers who raised this curved rampart of stone. On the dull water-logged days of spring perhaps only a mistle thrush has sung but I can also recall other days of bright sun when the dams looked less forbidding. I once was taking part in a film about the dams and on a bright morning was pointing out the carnivorous sundews and butterworts in a bog by the shore of the Pen-y-Garreg reservoir. Not far away a pair of lapwings was tumbling wildly in display over the rough grazings and a buzzard was hovering with slowly beating wings over a slope. A thin line of sheep, newly shorn from a farmhouse, began to unwind itself up the hillside forming, in turn, a pink fan across the rising slope. A curlew bubbled, and for me all was complete on that lovely spring day.

When I was a boy and first began to read about North Wales I discovered that there were three chief mountains – Plinlimmon, Cader Idris and Snowdon with its summit, Y Wyddfa, standing at 3,560 feet. March, when the gorse and *Daphne mezereon* are in bloom, is a fair month in which to see Plinlimmon. The route up is from the Aberystwyth to Llangurig road; it takes you to the highest summit of this mountain, which holds in its ample bosom the sources of the Rivers Severn, Wye and Rheidol. From the top there are breathtaking views to the long whale-back of Cader Idris, to Aran Fawddwy and even Snowdon itself. There are great sweeps of upland pasture rising high above the dark green carpets of Dovey and Tarenig Forest. Below to the north-west is the estuary of the Dovey, with its RSPB reserve comprising a region of ancient twisted oaks where tits, woodpeckers, redstarts, pied flycatchers, treecreepers and buzzard breed, of plantations of conifers with goldcrests and willow warblers, nightjars and grasshopper warblers and of marshland which forms a sanctuary for mallard, shelduck, redshank, common sandpipers, reed buntings, whinchats and sedge warblers. There are herons nesting in some Scots pines and an occasional pair of ravens.

Beyond the Dovey to the north rises the long igneous Ordovician wall of Cader Idris; firmly planted south of Dolgellau this ridge reveals to the naturalist some of the finest classical features of ice-shaped landscape – cwms, an almost circular corrie lake left when the glacier departed, and the U-shaped valley of Tal-y-Llyn to the south, comparable to the Pass of Llanberis in Snowdonia. At the end of that valley below Cader Idris is a lake rather like that at Bala, which can be found to the north-east of the mountain. It has a rare hybrid water lily and many kinds of fish but not the gwyniad – a white fish found only at Bala. The top of the escarpment of Cader Idris is a hard sill of igneous rock. The National Nature Reserve here includes almost a thousand acres of hill and lake and some fifty acres of foothill woodlands. In Cwm Cau the botanist who looks at the volcanic rocks and mudstones should find mountain sorrel, green spleenwort, roseroot, alpine scurvy grass, mossy and starry saxifrage and one plant that I am always glad to see – moschatel, with its four-sided green 'clocks'.

Cader Idris lies within the southern limits of the Snowdonia National Park – a region some fifty miles long and thirty-five miles wide. Large blocks of mountain and moorland are separated by deep and impressive valleys cut by ice and water and carrying the modern road system. Cambrian rocks occur here in North Wales, taking their name from the Roman name for Wales. At Harlech is a large oval area of Cambrian rocks known as the Harlech Dome, for the rocks are folded into a dome, sloping outwards in all directions. However, the Dome has been eroded and this shows the great layers of coarse sandstones and slates inside; the hard resistant grits have been weathered into barren, jagged mountains, whilst the slates form the lower countryside between. The upper part of the Cambrian succession consists largely of slates but the ranges of Snowdon, Cader Idris and Rhobell Fawr raise their Ordovician volcanic rock on three sides. Cambrian slates can be seen near Llanberis and at the great Penrhyn quarries at Bethesda, which once roofed much of the world; the imposing Siloam chapel in Bethesda built in 1872 was hung with green and purple patterned slates. There has been a long tradition of sheep farming, lead mining and slate quarrying in North Wales but once much of the area was well afforested with oak, birch and rowan growing rather as scrub, for the soils do not encourage the growth of large broad-leaved trees.

The soils and the climate do favour the growth of conifers, however, and so the Forestry Commission has turned to exotic trees for planting in the Cambrian State Forests: Norway spruce (the Christmas tree), Japanese larch with its golden russet autumn branches, Sitka spruce, Douglas fir, western hemlock, silver fir, western red cedar and lodgepole pine. The State Forests are extensive and in some ways impressive in their attempt to make up the shortage of commercial timber grown in this country, but these carpets of alien conifers in their settings do not have instant appeal and may indeed take away that sense of rolling space that the original denuded mountains possessed. Ystwyth Forest in the south has four thousand planted acres and some fine wooded gorges with oak, cherry and beech, all associated with Thomas Johnes who experimented with farming and sylviculture on the fringe of the moors. Myherin Forest is primarily a spruce, pine and larch forest set at over a thousand feet to the east of Devil's Bridge. A visitor coming by the Llangurig to Aberystwyth road will see the great plantings of Tarenig where trees clothe the clayey hillside. There is Hafren Forest in the shadow of Plinlimmon, where the River Severn rises, with its rows of Norway and Sitka spruce. Rheidol consists of four thousand acres of conifers, and larch and Douglas firs offer wooded views from the little narrow-gauge railway to Devil's Bridge. Coed Rheidol in its damp gorge is an oakwood National Nature Reserve, with Welsh poppies and many unusual ferns.

Not far from here is Cors Tregaron which is a raised bog – perhaps the finest in England and Wales – and a Nature Reserve with bog plants which are also the haunt of polecats, water voles, foxes, badgers and hares, while the breeding birds include several species of wading bird, buzzard and teal. In winter Cors Tregaron is a Wildfowl Refuge for white-fronted geese and many kinds of duck. I have known several hundred Greenland white-fronted geese to flight over this region of bogs, which serves as a roost and feeding zone, and also over the neighbouring hills. Mallard, teal and wigeon are common in winter and scarcer species appear too, such as shoveler, gadwall, pintail, pochard, tufted duck and goldeneye, while both whooper and Bewick's swans occur in many winters. The great bog of Tregaron is one of Wales's most exciting wildlife habitats. To the east of Talybont and between Aberystwyth and Machynlleth is Taliesin Forest, named from Tre Taliesin, the village called after the Welsh poet who it was said was washed ashore on the Dovey estuary; it embraces old oak coppices and high sheep walks planted with Norway and Sitka spruce.

Dovey Forest consists of scattered blocks and compartments of spruce, fir, larch, lodgepole pine, cedar and a magnificent stand of western hemlock, while for the historian there are such local attractions as the Roman fort of Maglona on the Dovey and Castell y Bere built by Llywelyn the Great. For the birdwatcher one of the local curiosities is Craig yr Aderyn near Towyn – the Rock of the Birds – a

133

precipice which is perhaps the only regular inland breeding site for the cormorant. Descriptions of this rock go back to 1682, and W. Catherall in 1828 called it 'a most picturesque and lofty rock' and went on to describe how 'large aquatic fowls from the neighbouring marsh may be seen majestically wending their way to this their place of nocturnal rest'. I have observed some twenty-odd nests on this rock, which can be seen from the road with good binoculars. One of the largest of the State Forests is Coed y Brenin, which embraces the woods around Dolgellau and is set about with fine mountains – Cader Idris, Aran Fawddwy, Rhobell Fawr and the Rhinogs – and the wide estuary of the Mawddach, filled with golden sand or pewter-coloured sea. The estuary and river are important wildlife sanctuaries, with orchids and rare plants, salt marshes and waders and wildfowl.

North-east are the high lands of Dduallt reached through hanging woods, with polecats, badgers and foxes, pied flycatchers, redstarts, wood warblers; the high zone is covered with hardy grasses and rushes. Conifers have been planted on many of the lower slopes and at the higher levels Sitka spruce and lodgepole pine compete with the more severe conditions on the hillsides. Coed y Brenin is a fascinating region and the dullness of the conifer plantations is counterbalanced to some extent by the old gold, copper and lead workings, waterfalls like Pistyll Cain and Rhaidr Mawddach, and Rhaidr Ddu – the last coming diagonally through a gorge into a dark, clear pool in the National Trust's woods at Ganllwyd not far from Coed y Brenin. Here the landscape varies from sea strand through broad-leaved forest and Forestry Commission plantation to moorland, hill grazing and rugged peaks. There is also Deudraeth Forest, which includes Hafod Fawr east of Ffestiniog and looks towards Trawsfynydd, where the first planting of trees by a Government agency in Britain for the production of timber took place.

When bare moorlands are planted with trees some interesting changes take place. The sheepwalks and hill grazings were once rich in mat grass or purple moorgrass, or gorse, heather and bilberry and populated with hill birds – meadow pipits, wheatears, ravens, ring ouzels, and so on; on the rather acid Rhinogs, for example, ling and heather are common with mossy saxifrage and bladder fern where red grouse and even merlins may nest. Perhaps there are a few pairs of dunlin around the peatbogs while golden plover and even dotterel are worth watching out for. In Snowdonia some of the bogs have been drained to accommodate trees, and such plants as bog orchid and alpine enchanter's nightshade have been lost. Before planting, the bare ground is often deep-ploughed by giant trench-diggers that turn over large furrows, break up the hard pans, drain the peaty land and improve the soil. In autumn after the ridges have dried out the tiny conifers are planted in the sides of the furrows. For the first few years some of the moorland birds hold on, particularly whinchats, skylarks, wheatears, meadow pipits and perhaps curlew, snipe and lapwing and even black grouse. I have seen short-eared owls hunting voles in Welsh plantations and there are often shrews, hares, foxes, polecats and perhaps the rare and attractive pine marten. As the conifers develop into the thicket stage with overlapping branches, willow warblers, grasshopper warblers, whitethroats, hedge sparrows, robins, wrens and yellowhammers establish themselves and after ten years blackbirds, song thrushes and chaffinches have joined the plantation bird community. I have seen bullfinches, redpolls and siskins in Welsh forests of this age. The black grouse has been scarce in Snowdonia but it has increased considerably in mid-Wales, favouring the hill plantations from their first establishment. When the trees are about fifteen years old their lower branches are cut away, a process called 'brashing'. The effect of this is immediate, since with the removal of the undergrowth the 'scrub' birds disappear and only woodland birds that can live in the canopy survive – magpie, jay, crow, woodpigeon, coal tit, goldcrest, chaffinch, sparrowhawk, owls, crows and sometimes ravens and buzzards. This kind of bird succession can be very well traced and studied in the woods of the Cambrian hills.

The Snowdonia National Park includes Snowdon – the highest summit in England and Wales and an extraordinarily wide variety of mountain and valley scenery, woodlands and streams. Three valleys run north – the Nant Ffrancon, the Pass of Llanberis and the valley from Beddgelert, and they all show the work of the ice sheets, with the weaker rocks being pulled away and the hollows filled with water – the Welsh llyns. In the extreme north is Conway with its own mountain, the celebrated Sychnant Pass and the Edwardian castle built by the military genius James of St George, which blossoms in spring with aubrietia, wallflowers and polyanthus and is commandeered by herring gulls and the jackdaws which I have watched tugging out lumps of dead grass from the masonry to build their nests in the round put-log holes which the thirteenth-century masons used to hold their spiral scaffolding. On Telford's bridge nearby gulls and street pigeons sit and sun themselves, while ravens sometimes call overhead and kestrels call to each other in their spring courtship. From Conway to Llanrwst mallard, lapwing, curlew and gulls can be seen scattered along the willow-lined valley, while many times I have seen the rooks cawing loudly as they passed over the Roman stronghold of Canovium on their way to roost. There are bats too at Llanrwst, just as that township's poet Evan Evans wrote about the school more than a hundred years ago

> Weaving their way of an evening
> Bats fly with silent beat,
> Where rang the poems of Homer
> And Virgil, elegant-sweet.

Beyond and to the south is Betwys-y-Coed inundated with conifers but with its famous Swallow Falls and in the Lledr Valley a mixed landscape of conifers and broad-leaved trees, which accounts for the many kinds of butterfly that occur in the district. There are silver-washed and dark green fritillaries, pearl-bordered and small pearl-bordered fritillaries, large skippers and green hairstreaks. The birds include black grouse, redpolls and siskins.

West of Conway green hills and scattered trees give way to more open moorland which is criss-crossed by stone walls; beyond the hill slopes stretch the smooth rolling hills of the Carneddau. In places the mountains fall deeply into Afon Dulyn and the Black Lake of Llyn Dulyn; here I have known the chickweed willowherb at its southernmost limit. Yet the hills and their summits here are soft and clothed with woolly hairmoss, reindeer moss, sedge, bilberry and crowberry; they unroll a long plateau where the skylarks and meadow pipits sing and the wheatears 'chack' at your approach. All around are wilder, more dramatic views of the harsher Glyders and Tryfan, formed from an acid rhyolite, where little vegetation grows. The Glyder range is a compact upland region. The spectacular cliffs of Cwm Idwal, known to mountaineers, rise under Glyder Fawr; they have been described as 'a hanging garden of rare Arctic/ alpines'; on these Ordovician volcanic rocks grow mountain spiderwort, least willow, mossy saxifrage, moss campion and other unusual plants. The romantically named Castell y Gwynt – the Castle of the Winds – rises as a shattered pyramid of rocks on the top of Glyder Fawr. The summit ridge is firm to walk on and consists of piles of boulders, rocks and scree, some strangely like ruined human constructions; here the sheep graze and the wrens sing. You can look across the Pass of Llanberis with its tiny winding ribbon of a road and lake with Arctic char, and the new construction works for the power station at Dinorwic.

Across the valley lies Snowdon – Eryri to the Welsh – The Rock of the Eagles. Snowdonia, as we have seen, is formed from a great variety of rocks, making a complex mountain land that has been folded and uplifted. Spectacular peaks, like those of Snowdon, with

136

ice-shaped summits, cliffs and pinnacle ridges are characteristic of igneous rocks, while lavas and ash are inter-bedded with sedimentary rocks – grits and mudstones. Subsequent movements of the earth and great pressure formed the slates whose excavation has scarred the countryside around Blaenau Ffestiniog, Dinorwic and Nantlle, and piled up enormous mountains of spoil. The quarrying museum near Llanberis I find a remarkable source of information and a tremendous help in re-creating the great days of the slate industry. The rocks belong to four main systems and are very old. Cambrian rocks disappear under those of the Ordovician of which most of Snowdonia is composed; these, in turn, slip under rocks of Silurian age, which form the eastern and southern flanks of the region. Because of the Ice Age there are few deep soils.

Snowdonia is compact and an evening's drive in the car, when the golden light shining low and revealingly across the landscape puts precipices in sharp relief, discloses fresh hollows and ridges and bathes the summits in glorious light, will unfold much of the region's geological history. From the Glyder range your eyes can wander over miles of moor and scree to Snowdon, Siabod, Arennig, Aran, Rhobell Fawr and Moelwyn, while Snowdon creates its own wreath of cloud across the valley. Here, wrote William Condry, 'you are shut in with the naked rocks of Glyder and your own thoughts about time and eternity and the slow forces that will weather away even these volcanic peaks at last'. As you descend perhaps a mountain fox, polecat or stoat may reward the journey, but the polecat is primarily nocturnal and sleeps up in his rocky den during the day. The lakes may have bog bean and water lobelia to ornament their waters while underneath trout, minnows and newts live their lives. Small heath butterflies flit across the stony path, while craneflies rise wearily up from the moss, and there is always the chance of hearing a raven or ring ouzel in song.

I have come to love Snowdonia and, in the words of Hilaire Belloc, have been moved 'with the awe and majesty of great things'. The proper route up Snowdon itself is by way of Pen y Pass and Crib Goch with its magnificent prospect, the Pyg Track, the Miners' Track by the old copper mine road, Lliwedd, the Watkin Path from Nant Gwynant, Yr Aran and Bwlch Main, Rhydd-ddu, the Snowdon Ranger from the Youth Hostel, the Llanberis Path along the railway line or via Cwm Glas. I well remember one July ascent by means of the puffing, toiling rack railway, opened in 1896, which finally disgorged a crowd of open-shirted and mini-skirted holidaymakers into a bitter, swirling mist at the summit station. Most of the visitors disappeared into the refuge of the hotel but a few clad in suitable clothes joined some of the hardier souls that had either walked up from Llanberis or clambered up one of the alternative routes, spurred on by sunny views of the cliff Clogwyn d'ur Arddu, with its northern rock cress, of Cwm Glas with its choughs, of the glacial lakes of Glaslyn and Llyn Llydaw with their saxifrage, moss campion, sandwort and alpine meadow rue and of the formidable red-brown ridge of Crib Goch.

I have cautiously explored the scree below the summit for fossils, particularly brachiopods, and found clumps here of woolly hair moss and starry and mossy saxifrage. In the mist I have heard ravens croaking while ghostly shapes flitted back and forth – the vague figures of scavenging herring gulls that in better weather solicit tit-bits from the visitors. By the summit cairn a gull will give the long call and be answered by another, both sounding strangely out of place in this high cold world. I have also been on the summit on days of outstanding clarity when I could look across Anglesey to Holyhead, to Yr Eifel on the Lleyn Peninsula, north-east to Carnedd Llewelyn and south-west across the Irish Sea to those mountains in County Wicklow that I described in Chapter 3. On such a day I once saw four mistle thrushes flying at a height of 3,400 feet towards the Pass of Llanberis, while some swifts were hunting two hundred feet below them. A raven was flying in the direction of Moelwyn with its veins of slate standing above Blaenau Ffestiniog, and I remembered how I had once gone down in a wooded cable truck deep into the mountain from which the slate was mined. As I came back into the daylight there were wrens singing among the heaps of slate spoil and nettles in the old workings. A badger had dug its sett into a high slate bank on the wall of the quarry, starlings were feeding young in a pile of slates, a couple of choughs called, their cries reflected from the stony wall, and a woodpigeon cooed from a small clump of grimy conifers. All round were monstrous barricades of slate slabs, adorned in places with a whole variety of ferns, and blue-grey in colour unlike the purple slate detritus in the Nantlle valley which towers over the swaths of spring daffodils.

Autumn and late winter in Snowdonia are for me often exquisite experiences. From the Snowdon Horseshoe Snowdon itself may be seen under snow, resembling a small white Matterhorn, cloud-bedecked and magnificent – a true and impressive crown to this mountainous region. In October near Llyn Padarn I have watched curlew, lapwing and gulls on the flat marshy tracts east of Cwm-y-Glo. The lower end of the llyn is overlooked by rugged ancient hills with smooth rocky outcrops or the mighty tiers and steps of the Dinorwic quarries, now noisy with heavy equipment and bright with

138

lights as the artificial lake above is finished to supply a hydro-electric power station in the valley.

One autumn day when low cloud hung in slow-moving grey-white shrouds above the mountains and old quarries, and the Pass of Llanberis lay before me, awesome and forbidding in pouring rain, I began the ascent of Snowdon from Pen-y-Pass. Great white cataracts of mountain water cascaded and tumbled down the rock faces. A buzzard circled above me looking larger than usual in the mist. Nine meadow pipits came in below the cloud base, flying south-west and trying unsuccessfully to find their way over a mighty cliff that lay across their migratory path. A wren sang a late song from the scree above me and two ravens cronked mournfully across the valley. Soon I was back in the familiar swirling mist. After I had descended I passed through Nant Gwynant, where dark green oakwoods rich in fungi, late golden gorse and bronze bracken fronds separated the lakes from the mountains on either side – Yr Aran on one side and Yr Arddu and Cnicht – one of my favourite mountains – on the other. On the shore of Llyn Dinas a cormorant was sitting like a disconsolate vulture, and further south I saw a heron beating on its broad wings through the Pass of Aberglaslyn – one of the celebrated beauty spots of Wales, with its bridge, cliffs, belts of Scots pine and river that after heavy autumn rains roars through in a veritable tumult of boiling brown, white-crested water. Just east of Tremadoc I paused to watch a sparrowhawk flying along the scarp of high cliffs, where red-helmeted climbers roped together scaled the dolerite faces like spiders. Coed Tremadoc is a National Nature Reserve with oakwoods growing on the steep cliffs and an interesting scrub community of plants; I have also seen ravens, jackdaws and buzzards here as well as red squirrels, and there may be pine martens too.

Nearby is the harbour of Portmadoc, with its almost mile-long embankment across Traeth Mawr, known as the Cob, built by William Maddocks and completed in 1811. Slate trucks came here from the quarries at Blaenau Ffestiniog, descending laden and running down by gravity. In the end truck of each train were the horses that were to pull the empty trucks back to the quarry. Steam engines were brought into use in the 1860s, and the recent revival of the Ffestiniog Railway has been a great success. Work is now going on to rebuild the line beyond Dduallt to Blaenau Ffestiniog. In 1971 I was employed by the BBC to present a television film about this miniature railway, which climbs up from the sea through farmlands, pastoral villages and the forested valley of the River Dwyryd to Tan-y-Bwlch. Happy Valley is clothed with sessile oakwoods where ferns grow on the trunks and rhododendrons rampage between the

139

boles. In spring and summer the woodlands echo with the songs of warblers, redstarts and flycatchers. There are magnificent scenic hill views between the trees and occasional glimpses of perhaps a buzzard, jay or woodpecker. I have seen herds of wild goats along the track; they also roam Cader Idris and the Rhinogs. Above Tan-y-Bwlch the scenery is more open and belongs to the sheep. Here they browse the mountainside and some die, to be picked clean by the ravens and crows, while on many days others run in panic before the puffing monster of the train or dash frantically up the hillside. There is a breathtaking view of the mountains from the line but I still find the sight of the Trawsfynydd nuclear power station unwelcome in the Cambrian hills.

Portmadoc lies below the barn-like summit of Moel Hebog, on which the River Dwyfor rises; it is a grassy mountain where bilberry, woolly hair moss and ferns grow. It looks down on spruce plantations south of Rhydd-Ddu and across Portmadoc to the Rhinogs, which lie behind Harlech Castle. These hills are rocky and littered with large stones, or covered with heather and bilberry where emperor moths can be found, and small lakes which abound with trout. There is a fine approach from Llyn Cwmbychan through woods where warblers and pied flycatchers sing their spring paeans of passion, to the bare mountain tops with many stone circles, dolmens and burial mounds, reflecting the extensive human occupation three thousand to four thousand years ago. The hard Cambrian rock does not have many plant species. At the lower levels there are foxes, badgers, polecats, wild goats and hares. The lesser twayblade orchid grows among heather and bog moss, and on the highest parts of the range there may be golden plover and dunlin. To the east lie Rhobell Fawr and the two peaks of Arennig, ashy and formed from lava flows, where the curlews bubble, whinchats and wheatears give their pebbly calls and ring ouzels sing their wild pipes. Perhaps it is around here that a pine marten might show itself or a dotterel trip away over the rough, stony and fissured ground.

The finest views of Snowdonia are those from the island of Anglesey – flat until it ends in the recumbent lion of Holyhead Mountain with its maritime heath and seabird cliffs – and from the Lleyn Peninsula which has been a favourite holiday haunt of mine for years. Not only is Lleyn an Area of Outstanding Natural Beauty but it is also a fine centre from which to explore North Wales. Here the old county of Caernarvonshire – now simply a part of Gwynedd – points a drooping and despondent finger down in the direction of Southern Ireland. The Peninsula is a long arm of land dotted with hills of igneous rock that look down on sea cliffs and bays, gentle

140

farming country with white or grey houses, straight main roads and countless miles of narrow winding lanes. At Criccieth is the ruin of one of the Welsh princes' castles, at Llanystumdwy the early home of Lloyd George, at Pwllheli sea holly and a large holiday camp. Aberdaron is the last resting place for the Bardsey pilgrims, Plas-yn-Rhiw a National Trust garden, Abersoch a yachtsman's paradise and off the coast there are St Tudwal's Islands.

My base has always been at Nefyn on the northern shore; it was here in 1284 that Edward I held a tournament to mark his triumph in Wales. In spring gales come rushing in on the heels of mist, rain and sleet. Across Cardigan Bay the mountains of Snowdonia and mid Wales are often shrouded in deep snow and wreathed with ominous black clouds. Herring gulls wail antiphonally from the rooftops, jackdaws call from the chimneypots in the village and collared doves moan from the isolated clumps of pine and cypress. At this time of year the roads and lanes are almost free of traffic so that one can look up at the passing buzzard or raven. The roadsides are alight with golden gorse, and the thin bleat of lambs can be heard everywhere. Celandines and violets star the grassy banks and primroses smile up from among the fleshy discs of pennywort. Viewed from Snowdon the Lleyn Peninsula looks like a chain of conical hills. In fact, it is made up of a number of isolated mountains ending with the steep bluff of Ynys Enlli, or Bardsey Island – the 'Island in the Current' – with its twenty thousand sleeping saints. It is an island of poor pasture, cultivated land, and its mountain Myndd Enlli, a bold and striking ridge of rock, clothed in fescue, heather, bracken, gorse and sea pink. Many small birds nest on the island. I have seen choughs crossing the two miles of sea to the Peninsula's end at Mynydd Mawr. There are rare breeding birds too – perhaps storm petrels in fissures in the rocks and Manx shearwaters which flight at night to avoid the attentions of the gulls – *'âmes damnées'* of the Bretons, whose strange calls and caterwauls can be heard enlivening the nocturnal gloom. Bardsey too has grey seals, lizards and slow worms.

Back on the Peninsula is the ridge of Mnyndd Rhiw, where I have explored the remains of the prehistoric axe factory under a chorus of singing skylarks and meadow pipits on the mountain slope, parachuting wheatears and bleating hobbled sheep. I have climbed Garn Bodvan above Nefyn a number of times, reaching its stone-walled fort enclosed within its Iron Age ramparts. In summer as one climbs up through serried ranks of conifers, goldcrests and coal tits sing and provide company on the way, green woodpeckers yaffle, and sparrowhawks soar overhead on broad-fingered pinions. Around the summit cairn hoarse-voiced ravens croak and tumble in

141

the stiff sea breeze. To the south-west is Garn Fadron – rockier, wilder and more severe, and adders bask on the southern slopes in the spring and summer warmth.

Among these hills, narrow lanes, fields where shepherds train their dogs and work them too, and tiny villages, there are relics of an ancient race – master-builders in stone who left behind standing stones, cromlechs, burial cairns and hill forts. My favourite ancient site is, however, that of Tre'r Ceiri – the Town of the Giants – standing at 1591 feet on one of three cone-shaped peaks of Yr Eifel, or The Rivals, which guard the northern entrance to the Peninsula and the old pilgrim's track from Caernarvon to Bardsey Island. Here two huge stone defence walls cast their shielding arms around the remains of more than fifty-one stone-walled huts, some of which are round, some oval, others oblong and some semi-detached. Now the only inhabitants of these fallen homes and standing walls, dating perhaps from AD 150–400, are the wrens that sing their passionate notes among the stones. This may have been a summer settlement when the improved weather allowed pasturing on the hill and the easier pursuit of war.

The seaward member of the three Rivals is a mountain of pink and blue-grey rock cut into giant steps by the workers from nearby Trefor. One flank falls in a giant precipice into the sea and provides footholds for seabirds, and I have seen peregrines displaying at eye-level along its face. At its foot on the southern side in a remote little cleft known as Vortigern's Valley, where traditionally the King of the Britons fled to die after betraying his country to the Saxons, lies the deserted village of Porth-y-Nant. Two terraces of grey and eyeless partially tiled houses depart at right angles and form two sides of a square; on the other two sides are the manager's derelict house and the chapel. There are clumps of sycamore, straggling privet hedges and straggly roses. A steep path leads down to this abandoned quarry village, once a hippy colony but now inhabited by sheep, large black goats and the ubiquitous wrens. The deep croak of ravens adds to the sense of gloom that hangs like a pall over this deserted, decaying village and, briefly, as a distant bird sings above the gentle breathing of the sea, there is the momentary whisper of a ghostly Welsh voice! Now it may become a Welsh Language Centre!

The steep hillsides above the village echo with the gruff 'kyaaas' of glossy blue-black choughs with coral legs and bills. I have watched as many as twenty-four of these intriguing birds going to roost on a cliff face above the village. In spring snow comes blundering into Porth-y-Nant out of skies jet-black with ill-omen and I take shelter in the chapel. Moments later the sun comes streaming out, lighting up

Anglesey and the chaffinches arriving back from winter in Ireland and dipping in from the sea. All three kinds of British diver are fishing in the grey and steely water below the cliffs. A chiffchaff begins to sing and in its repeated phrases lies the promise of the coming summer when the dorbeetles emerge and devilsbit, kidney vetch, wild pansy and swaths of montbretia clothe the cliffs. In front of the ruined post box at Porth-y-Nant a wheatear bobs; just back from equatorial Africa he brings a returning sense of good cheer to this remote spot in the Cambrian hills.

10. From Dee to Solway

North from Wales the coast of England runs in broad peninsulas and bays to the Borders and Southern Uplands. At irregular intervals it is deeply indented by the estuaries of rivers – the Dee, Mersey, Ribble, Lune and Duddon – and great inlets like those of Morecambe Bay and the Solway Firth. Inland at first the countryside is flat in the Cheshire and Lancashire Plains but to the east rise the grey hills of the Pennines and to the north the dark purple mountains of the Lake District. The coastal strip embraces some of the most fascinating and important wildlife refuges not only in Britain but also in Western Europe; one bay is the most valuable for wintering wading birds in the whole country and indeed much of the emphasis of this chapter will be towards wide seascapes, bleak open mudflats where wildfowl and wading birds gather in thousands under immense and often beautiful skies.

Cheshire as a county can claim only a short coastline but it boasts an estuary and three small islands with almost international fame – those of Hilbre. The estuary is that of the River Dee where Charles Kingsley recorded

> The creeping tide came up along the sand,
> And o'er and o'er the sand,
> And round and round the sand,
> As far as eye could see.

The Dee is famous for its salmon that leave the sea and swim upriver towards their upland spawning grounds between the months of February and August, but the estuary is a strange, sometimes enticing, sometimes forbidding world of ever-changing wind and tide. One moment it is still – at peace – and then a change of wind to the north-west rolls the water in at a frightening, breathtaking speed or spreads a thick blanket of dank fog. The Hilbre Islands are outcrops of sandstone exposed at half tide. The months of autumn and winter bring enormous numbers of birds to the area – a few geese, many ducks and countless hordes of wading birds down from their nesting grounds in the Arctic. These are also joined by our own nesting waders – curlew, oystercatchers, redshank, dunlin and ringed plover. At dead low tide some fifty square miles of mud and sand are exposed to provide the marine life that supports the wading birds. The highest tides cover all but the Hilbre Islands, so that huge flocks of birds are forced to gather on the only undisturbed land above water.

My first visit to Hilbre was by pony and trap across the sands from West Kirby, past the high rippled sand ridges, through the shallow runnels of water and finally along two deep ruts in the solid rock that climbed up to the top of the island. Oystercatchers were probing in the mud and redshank, dunlin and a grey plover were huddled together in a shallow pool. A vast thumb-print in the sky revealed itself as a pack of several thousand knot, which poured down on the flats like a giant waterspout. On several March days I have concealed myself in a hide some two hours before high water and just above the tide mark to wait for the wading birds to land on the bare rocks around me. Inexorably the tide comes up and then suddenly there is a mighty roar of wings and thousands of knot begin to swirl overhead to pitch and pack in a grey mass of birds that spreads around three sides of the hide. They are joined by hundreds of dunlins, oystercatchers, redshanks and smaller numbers of turnstones, purple sandpipers and an occasional herring gull or godwit. As I have sat in the hide I have recorded on tape the cacophony of wild calls, noting in my diary 'the low purring notes of massed knot, the clear liquid whistles of curlew and redshank, the staccato trilling of turnstones, the high nasal twitter of dunlin and, above all, the shrill piping of the oystercatchers'. There have been up to five thousand knot within thirty feet of my hiding place! The birds eventually fall asleep with their heads pushed under their scapular feathers. As the waters begin to recede the great flock begins to melt away. I have sat amongst these waders in warm sun and in snow and find the experience one of the most thrilling in my life as a naturalist. Many times over the Dee I have watched and wondered at the packs of knot like rainclouds in the sky which bank in the sunlight to glow gold for a second, and then as they turn almost disappear before showing up once more as an oval smear against the sky – intricate manoeuvres performed in perfect unison.

A bird-ringing station has been established on Hilbre and it is also

145

a local nature reserve. Its rarer bird visitors have included both marsh and Montagu's harriers and short-toed lark. I found sea wormwood and sea spleenwort growing there, and the late Norman Ellison showed me where some clumps of primrose peerless were growing in one of the ponds. The waters of the Dee hold flatfish, shrimps, lumpsuckers and occasional lesser octopuses and porpoises. Some years ago I went out from Caldy with a film unit in an inshore shrimp trawler, or 'nobbie', to trawl for shrimps in the shallow waters around Hilbre. The boat was half-decked, cutter-rigged and equipped with a steaming cauldron in which to boil the shrimps. From the 'nobbie' I had fine views of the estuary and the two strips of land along each bank. Grey seals sometimes lie up off the West Hoyle Bank and there are hedgehogs on the largest island. The breeding of a few pairs of house sparrows has revealed their dependence on man and his animals – horses, goats and poultry – for regular food, and a need for minor invasions from the mainland to sustain such a colony.

Lying between the broad waters of the Dee and the Mersey is the Wirral Peninsula. It is a flattish plateau of red Triassic sandstone, crossed in places by green lanes and bounded on the east by the industrial complexes and dormitory towns of Birkenhead and Wallasey. I have become very familiar with the western shore from Red Rocks near Hoylake down to Barton. The Dee has silted up and since the first discovery of cordgrass growing in the mud off Parkgate about 1944 a thick mat of vegetation has colonized the shallow shore. I usually stay at Parkgate, where during the war I saw boats moored at the quay. It is a pleasant amalgam of varied houses, narrow passages and reminders of an active past. A busy eighteenth-century port and then a fashionable resort, it finally died as the River Dee silted up. Now it is frenetically alive with cars and visitors at the weekend but for the rest of the week it is a quiet place for birds and birdwatchers. Acres of cordgrass, deeply rooted in the mud, now stretch like a grey-green meadow from the seawall half a mile and more into the Dee. In autumn there are pools where I have seen greenshank come to feed and wrens forage in the vegetation. In spring a mistle thrush sings on a wind-tossed bough near the quay – a sound that carries me forward in spirit to the time 'when the blossomy boughs of April in laughter shake'.

Across the Wirral on its other shore the River Mersey presents a different and less edifying picture. As long ago as 1207 King John saw the possibility of a port on the river and now Liverpool boasts seven miles of river wall and nearly forty miles of quay. I have driven from the peace of the Wirral through the Old Mersey Tunnel into

Liverpool's dockland to see the street pigeons and watch great mechanical scoops ladling out raw brown sugar and a New Zealand boat discharging 161,384 frozen lamb carcases which dropped like leaden weights onto the conveyor belt below. From the basin nearby came the urgent calls of ships anxious to break out through the huge lock into the open river. Even in these apparently unpromising conditions there are birds to see besides the street pigeons and sparrows. I have watched mallard, tufted ducks, crows and gulls, and from the Isle of Man steamer pushing up through the Mersey's streaky, befouled water I have seen common terns fishing off the bows. Some counts on the Mersey have shown up to five thousand mallard and seven thousand pintail, gathering on the water.

Inland from the Wirral lies the Cheshire Plain – my father's home and that of his forbears. About two hundred million years ago in Triassic times the region lay under an ancient, inland sea which in very hot, dry conditions gradually dried up but was refilled from time to time. Tides and wind brought the sea back across ridges and low dunes to fill the salt lakes again – lakes that stretched from Ireland to Russia and beyond. As a result layers of salt formed many hundreds of feet thick and these, in turn, became topped by later marls and glacial drift. Before man began to interfere with these layers above the salt, land water drained down to it and produced streams of salt brine which came up to the surface. Cheshire salt was exploited by the Romans and today the vacuum plants and tall pencil-like chimneys of the saltworks dominate the landscape around Northwich and Middlewich.

The pumping of brine and salt-mining have caused the land in places to sink and form 'flashes'. The mining of rock salt since the 1750s contributed to the collapse of the workings and water dissolved many of the columns of salt left to hold up the roof of the tunnels. Many of the flashes filled with water and for me their attraction are the great crested and little grebes, redshanks and reed buntings, kingfishers, herons, mute swans and Canada geese which come to them. Yellow wagtails sometimes nest around the flashes but some have become filled with industrial lime waste and only gulls seem to survive in this man-made habitat! In the season of migration waders sometimes drop in to feed and rest. This part of England is composed of gently rolling countryside, small woods, mosslands, fens and meres often with reed swamps, streams and canals. Drinking pools for cattle have been formed from diggings for marl to improve the quality of the land; these marl pits are often fringed with sedges, yellow iris, and purple loosestrife which, in turn, provides sanctuary for moorhens, reed buntings, sedge warblers,

147

robins and hedgesparrows and shelter for mallard and teal.

In winter the Cheshire days are sometimes misty and depressing but others are born still and warm in a pale sun. Black-headed and herring gulls from the Mersey stream into one flash that I know to bathe and rest, while common gulls assemble on the new plough-lands near Great Budworth or forage and scream along the edge of Marbury Mere. In November new flights of mallard come to the meres and small parties of tree sparrows sit about in the hedges and chirp spasmodically. Witton Flashes sometimes hold numbers of dunlin. Waders, terns, ducks and grebes rejoice in the Cheshire meres, set in the rich farmland, hedges and fox coverts of the county. To stand among the reeds is to share with Charles Kingsley the 'joyous clamour from the wild-fowl on the mere'.

I have often visited Rode Heath, where my father was born in 1872. Here the Trent and Mersey Canal to Runcorn runs past a red brick wall – part of the old Lawton saltworks closed when the Salt Union was formed. Now there is a flash where the works collapsed into a deep hole. My father's father was a boatman working boats to Runcorn and carrying salt blocks for shipment to Africa and South America. My father was born in a cottage near both the canal and the saltworks and his diary sheds an interesting light on the Cheshire salt industry.

There is in the village of Lawton a small salt works. These old works, tradition said, were worked by the Romans. The source of the brine for the Lawton Salt Works and the works in Old Rode, was a well in Lawton Parish from which the brine was pumped by an undershot water-wheel, still there in ruinous condition when I was a boy. It was a fascinating game to climb over the old paddles.

The brine was pumped into two reservoirs through hollow tree boles and run into large iron pans. Underneath long flues extended to the end of the building; through these the heat evaporated the water from the salt. Long scrapers kept the collected salt at the bottom of the pans on the move. Along the sides were stacked perforated wooden moulds. When the salt had crystallized it was lifted by perforated scoops and packed in the moulds. When it had sufficiently drained, the salt was carried into the drying chambers and the moulds emptied, resulting in the familiar shaped blocks we see in the shops. These were for household use. The salt crystals were not put into moulds but emptied into heaps for cheese and other manufacturing purposes.

My father was expert in making skeleton ships, fully rigged. The curved hull was made from thin wood split from the round sides of cheese boxes. The ships were dipped in the boiling salt brine, and they were frosted with a granular coating of dried salt. One of these in a glass case was to my childish fancy a wonderful work of art.

148

Salt is still an important industry in Cheshire and the methods, as I came to find out when I was involved in filming in a modern works, are only updated; instead of the brine being boiled in open pans it is now evaporated in large enclosed cylinders. My descent into the Meadow Bank Mine at Winsford was enthralling, since it is the only working salt mine in Britain. It was first worked in the nineteenth century by the light of tallow candles and with black powder, pick and shovel. Now mechanized systems of extraction give a production capacity of 1,300,000 tons a year. Vast crystalline tunnels, caves and walls glisten amber and liquid brown in the powerful lights. Much of the mined rock salt is used for clearing ice and snow on our roads but some is used as a fluxing agent for metal refining, for glazing drainpipes and other clayware and as a sodium fertilizer. As I wandered through these caverns I thought how much unseen and perhaps unknown goes on under the inquiring naturalist's or observer's feet.

One way of travelling north from Cheshire and Lancashire is by ferry to the Isle of Man. One crossing I made was enlivened for me by a party of Mancunian ornithologists in anoraks, armed with field glasses, clustered forward in a loose gaggle and vying gently with each other to spot the first kittiwake, first skua and first auk on the murky waters of the Irish Sea. From a less exposed point of vantage I logged four Arctic skuas and a great skua as well as Manx shearwaters, gannets and other seafowl. After the four-hour crossing from Liverpool my plan for traversing part of the island is to take a horse tram along the elegant two miles of promenade to Derby Castle, before ascending an old trailer coach of the Manx Electric Railway for Ramsey. This journey is a sheer delight, running along the edge of huge seacliffs, through cuttings, across farming country and over hills. There are hedges of fuchsia and escallonia in blood-red flower and I once saw a Manx cat walking on long legs, almost self-consciously round a corner of Laxey Church. Perched on the wooden seat across the open-sided coach one can see all around, and I remember one September visit watching magpies and collared doves, hooded crows and jackdaws in the fields along the line. Many migrant wheatears were standing about on garden lawns and even front drives. Rabbits were everywhere and I watched a white cat stalking a jet black rabbit in a field. Ramsey in its sheltered sunny bay was echoing with the autumn songs of robins.

In some ways the Isle of Man is like a miniature Lake District with slates of Upper Cambrian or Ordovician age forming Snaefell and the moorlands, with Carboniferous rocks in the north and south and some New Red Sandstone. The island was glaciated but is rather poor in its post-Glacial fauna, suggesting that its link with the mainland was in Glacial times. It lacks certain mainland species such as the tawny owl but has long-eared and short-eared owls. Stonechats, wheatears, meadow pipits and grasshopper warblers are widely distributed and there are choughs as well as many kinds of seabird including black guillemots. The colony of mysterious nocturnal Manx shearwaters on the 616-acre cliff-bound rock island of the Calf of Man appears to have been there since 1014 according to the Icelandic *Njal's Saga*. Notable absentees from the fauna are the mole, common shrew, squirrels, snakes, slow-worm, newts and common and natterjack toads. It is interesting to note that natterjacks can be found in the barren dune slacks on the nearby Lancashire coast. There is an area of pasture and dunes at the Ayres, which in plants and insects is quite rewarding.

Lancashire on the mainland is a curious mixture of textile towns, huge conurbations, extensive farmlands and high sheep grazings. For hundreds of years the county was famous for its marshes and mosslands, which Michael Drayton called 'low watery lands and moory mosses'. Many of these mosses had been drained by AD 1800 but regions such as Chat Moss, Simonswood and the marshes near Southport with their scrub and swamp defy taming by man and remain sanctuaries for wildlife. Martin Mere is a piece of fenny lowland formed from artificially drained peat. Most of the original mere has been drained but the Wildfowl Trust has purchased 363 acres of land. In winter many hundreds of pink-footed geese come to the Reserve and there are perhaps 26,000 birds in this part of Lancashire. Pintail, teal and snipe occur in their thousands while peregrines, hen and marsh harriers and wild swans have been seen as well.

I have stayed at Martin Mere in December and February to look at the Trust's collection of wildfowl and at the wild birds that come to feed from the surrounding area. One cold February evening I watched a party of a hundred ruffs assembling on a watery bank; snipe were probing in the frosty ground and skeins of wild pink-footed geese began to pour into the mere for the night. I had laid a microphone on the edge of the water to record their resonant 'unk-unks' and high 'wink-winks'. The next morning I was up at half past four to record the harsh quacks of the mallard, the slurred whistles of drake wigeon, the low penetrating 'erk-erks' of gadwall and the pops and explosive yells of coot. There were also the conversational calls of the geese. For an hour and fifty minutes I waited for the pink-feet to leave the mere and go off into the open country or to the Ribble Estuary. At exactly ten minutes to seven the first goose put up its head in anticipation. I switched on the recorder and suddenly these attractive and musical birds left their partly frozen lake, calling and thrashing the air with their wings, as they flew off into the rose-pink of a new and infinitely beautiful dawn.

The Ribble estuary lies between Southport and Blackpool and has been nominated as the fourth most important estuary for wildfowl and wading birds in Europe. There are extensive salt marshes at Cossens Marsh, Banks Marsh, Hesketh Out Marsh, Longton, Clifton and Warton. Grassy walls and embankments cut off areas of mud, marsh and channels from the land. The summer vegetation is largely red fescue, sea meadow grass, oraches, common seablite and sea pink and some of the region is grazed by cattle. The natural salt marshes are alive in summer with breeding mallard, shelduck, redshank, oystercatchers, about ten thousand pairs of black-headed gulls and a thousand pairs of common terns. Skylarks sing all day high above the meadow pipits with their repeated chipping notes and the yellow wagtails, bright canary yellow and calling 'tsweep', as they dash about on the ground searching for insects. On the reclaimed and poorly drained marshes there are more larks as well as lapwing, snipe and redshank and on the better drained land one can see corn and reed buntings and partridges. The Ribble is best known in winter, however, as a base for wading birds – perhaps the estuary holds a quarter of all the British sanderling and black-tailed godwits, one fifth of all the knot, bar-tailed godwits and grey plover, and one tenth of the dunlin. The roost on Crossens Marsh is a truly remarkable sight since it can hold up to 150,000 waders. It is not impossible to see 100,000 knot and many thousands of pink-footed geese, while there is now a flock of a hundred Bewick's swans in most winters. The Ribble forms part of a complex of estuaries in the Irish Sea – the Dee, Morecambe Bay and the Solway – and itself is one of England's greatest wildlife treasures. The Nature Conservancy Council's decision to purchase the area to prevent its being privately reclaimed is therefore most welcome. 'Reclaimed' is in the nature of a misnomer; it should be 'claimed', since we did not possess it in the first place!

Whenever I visit the Lancashire coast and the Ribble I also travel north to Morecambe Bay, where again vast expanses of tidal flat and marsh unfold under the lights of Morecambe, stretching like a grey and green carpet towards the Lakeland fells and hills. Here the RSPB have bought more than six thousand acres of the Bay as a reserve, and it is without doubt the most important region for waders in Britain and one of the most important in Europe. Here, as in the estuary of the Ribble, the rich invertebrate life and the undisturbed regions for feeding and roosting attract large numbers of birds.

151

Again red fescue and sea meadow grass with white bent and sea pink are the commonest plants, but I have seen sea aster and scurvy grass as well. The breeding birds are similar to those of the Ribble but the old sea wall is often used by wheatears, pied wagtails and little owls for nesting. Here are many wildfowl and waders in winter and the seasons of migration. Cast your binoculars over the marsh near Hest Bank and you can watch thousands of dunlin packed in grey and white strands along the water's edge, and there are curlew, knot and other waders here too. In spring the sprightly ringed plovers running on the shore may be on their way to Greenland. The flats are rich in bivalves, shrimps, marine worms and sandhoppers, which all help to feed the countless birds. In 1962 a fish trap was in operation on Morecambe Bay set amongst great mussel beds – 'skeer' is the local name – which had been continuously operated by the same family for five hundred years; its catch included flounders and plaice, eels and mullet.

There is a threat to the Bay since plans have been put forward to create a barrage across it. Such a development would alter irrevocably the wide and unique panorama of grey-blue water, green marsh riddled with small pools and criss-crossed with meandering watercourses, where one can hear the mutter of godwits, the incipient bubbling of curlews, the honk of greylag geese, the whistle of wigeon, and the laugh of shelduck coming clearly across a scene, part land, part sea, beyond which the golden sun of autumn drops in its shallow arc into night's oblivion.

Many of the wildfowl in Morecambe Bay fly regularly to the RSPB's fenland reserve at Leighton Moss, close to the north-east tip of the Bay. It is situated in a pleasant wooded valley among limestone hills near the village of Silverdale. There are mature woods, reedbeds and three main and several smaller meres. It is the reedbeds with their flourishing reedmace and yellow iris which I have found the most fascinating part of the reserve. On the dry fen edge it is possible to find panicled sedge, meadowsweet and fen orchid. The great attraction of the reedbeds are some nine pairs of bitterns, which are here at their most northerly breeding station in Britain. They boom in their lugubrious bovine way from mid-January to June and seem to be particularly stimulated by the sound of diesel locomotive hooters on the nearby railway line. With the bitterns there are reed warblers which chatter through the summer days; water rails squeal and grunt and there are mallard, teal, shoveler, pochard, tufted duck and occasionally garganey. Coot and moorhens are abundant and the rare spotted crake may breed as well. There are also sedge warblers and reed buntings. Many duck come in the winter months and grey geese can be seen flying over the reserve.

Otters are regular at Leighton Moss – five were seen in one week when I was there – and the meres hold eels, rudd, perch and pike. Red deer sometimes wander onto the reserve as well. Nearly three hundred different species of moth have been reported, including five kinds of hawkmoth, while the fen wainscot and the bulrush wainscot are dependent on the reeds and reedmace. Peacock butterflies are a feature of the grassy walkways and visit the flowering hemp agrimony. The highlight, however, of a visit in December was for me a party of waxwings. All my life I had waited to see this rare invader from the Continent, missing birds in Holt, Norfolk, by five minutes, in Rugby by ten minutes – the list of failures was enormous. There were seventeen at Leighton Moss sitting in an ash tree as I interviewed the warden, John Wilson, for my Radio 4 'Countryside' programme. From time to time these grey and chestnut birds with their attractive crests flew down to feed in some hawthorns or to land in the roadway some twenty feet away from me or pitch in a field to drink from puddles in the earth. And then there were those high weak trills – notes that I had never before heard in the wild. The next waxwing that I saw was near Peterhead two winters later and the species was no longer my elusive bird.

That same year in which I saw the first waxwings I also made a sentimental journey back to the estuary of the River Kent just to the north – a spot that I had not seen since the last war, when I was at the RAF station at Millom. I arrived at Sandside on a cold winter's afternoon. The Lakes mountains stood sharply etched in black and white. A vast area of sand, marked like a tapestry with watery runnels, stretched away from me towards a distant wooded shore. From the sandbanks rose a large flock of rooks and jackdaws. A red-breasted merganser was fishing close by in a creek, cormorants were labouring through the sky overhead and curlew, redshank, ringed plover, dunlin and many gulls were foraging on the sandy ridges. As I walked over the damp and slightly yielding sand, squeezing out dry patches around my shoes, I could hear above the calls of the wading birds and gulls the deep notes of greylag geese out on the sound. Just around the corner is Grange-over-Sands. At one time the traveller left Lancashire from Hest Bank to journey across the sands to a point on the Cartmel coast. Ann Radcliffe described in the eighteenth century how she made that crossing: 'the tide was ebbing fast from our wheels, and its low murmur was interrupted, first, only by the shrill small cry of seagulls, unseen, whose hovering flight could be traced by the sound, near an island that began to

153

dawn through the mist, and, then, by the hoarser croaking of sea-geese'. This was then the shortest, albeit most hazardous route across Morecambe Bay, and Wordsworth called it 'a decided proof of taste' to enjoy the views of the wooded slopes and mountains beyond. The little grey town of Flookburgh looks across a mile or so to the sands of the Bay and it rests, silted up in its corner, against the wooded foothills of Lakeland and the impressive shapes of Coniston Old Man, Witherlam and Dow Crag. From here the 'cart-shankers' used to follow the tide out for shrimps, each driving a tall horse which pulled a two-wheeled cart. I can remember twenty years ago how they also took with them a 'jumbo' or wooden plank that the fisherman rocked back and forth on the sand to encourage the cockles to come to the surface, where they were collected up with rakes. Further still, around this tortuous coast is Walney Island, with England's most southerly breeding eiders and huge colonies of herring gulls and at least one third of the British and Irish total of breeding lesser black-backed gulls. It lies just off the mouth of the Duddon estuary:

> Wide stretch broad Duddon's sands,
> Guarded by Black Combe
> That threatens a summer sky
> Of waterlogged and grey compose;
> The sea, a thin far line of white,
> Curved in convexity
> From Ravenglass to the white pall of Barrow.

Near the sea is an area of boggy pool and sedge, close to the old air-field from which I flew. Here cock reed buntings sing in the grassy, sedgy growth and sedge warblers utter their chattering whistling medleys, full of verve and apparent inconsequence. At Silecroft in the shadow of Black Combe there are belts of scrubby thorn full of linnets, greenfinches and redpolls. The pink-breasted, red-chinned redpolls have become common in Cumbria and I often listen to them flying overhead and giving their high metallic flight calls or brief wheezing songs. Not far from Duddon Bridge is a wood – damp and verdant after rain, full of ash trees and hazels with a field layer of acrid ramsons, wild strawberries and forget-me-nots; here willow warblers and blackcaps sing in May, the soft silvery cadences of the former being punctuated by the liquid warbles of the blackcaps. From here the River Duddon reaches back towards its source – back through Ulpha, under Harter Fell and up to the top of Wrynose Pass, where it rises near Three Shires Stone. This is an attractive

154

way into the Lake District and one which inspired Wordsworth to write many of his sonnets.

About one tenth of the total area of England and Wales is devoted to National Parks – that is, land selected by what is now the Countryside Commission 'for the purpose of preserving and enhancing their natural beauty and promoting their enjoyment by the public'. We have already looked at Snowdonia, Dartmoor, Exmoor, the Brecon Beacons and the coast of Pembrokeshire. I have a very special regard too for the 866 square miles of the Lake District National Park, with its magnificent mountain scenery, woods of oak, ash and birch and many lakes and tarns. England's highest mountains – Sca Fell Pike, which like much of the Lake District is owned by the National Trust, Helvellyn and Skiddaw look down on lakes which include England's largest – Windermere. Sculptured craggy volcanic peaks of Ordovician age like Helvellyn and the Langdale Pikes, Sca Fell, Great Gable and Dow Crag rise up above Silurian moorlands.

The northern part of the Lake District is formed from the Ordovician Skiddaw slates laid down in an ancient sea from Ennerdale by Crummock to Derwentwater, Skiddaw and Saddleback. These are rather softer hills but Sharp Edge is a craggy outcrop. The Silurian slates provide hills and valleys with a gentle, rounded appearance. Both Ordovician and Silurian rocks were inundated by a Carboniferous sea some 280 million years ago and were then uplifted. The Carboniferous or mountain limestone capping of the Lake District was eroded away by wind and water and now only appears on the fringe of Lakeland in the northern Pennine range from the Craven Uplands to Cross Fell. Then the ice sheets put a finishing touch to the landscape, cutting into and plucking out material from the valley sides to form the crags of Sca Fell, and Dow, leaving moraines and lakes, *roches moutonnées* and waterfalls. There is almost infinite variety in the Lakeland scenery and each of the chief dales has its own character – Wastwater stony and arid, intimate yet windy Grasmere, Esthwaite Water shallow and graceful, Elter Water and Coniston wooded and sedgy, Thirlmere set among serried ranks of conifers, busy, touristy Windermere and outlying Ullswater – a gem on an early summer morning.

These were the chosen landscapes for Ruskin, Wordsworth, Coleridge and Southey and after experiencing spring in the Lakes with the glory of the daffodils, the purple flush of the birches, the still, mirror-like lakes and the windflowers and celandines along the banks of tumbling rushing becks, I could understand why! I have stayed at Skelwith Bridge below Elter Water, where the river forces

155

its way through a narrow gap and then rushes under a stone bridge, where the midges dance and the trout lie deep and still. In spring the buds on the sycamore stand up fat and waxen, like tiny yellow candles, and the ash buds swell black and golden. Red-breasted mergansers fly along the river and lapwings tumble and call in their spring ecstasy. One day in a nearby field half a dozen jackdaws queued up to take their turns at pulling out lumps of wool for building their nests from the tail of a recumbent sleeping sheep. Thrushes and tits sang their spring melodies and a grey wagtail spent most of his time on the water-washed boulders in the beck, flying up every now and then to sit in a willow gay with yellow catkins. At Tarn Hows, where naturalized larches raise their pyramids around a somewhat artificial but appealing tarn, the chaffinches flock in from all directions to be fed. Some of them will sit boldly on the roof or bonnet of the car and even on the wing mirrors. Chaffinches can also be heard singing in the larches, while willow warblers pour out their incessant silvery notes and coal tits chime away in the high canopies of the trees. From the water itself come the sharp calls of coot and the screams of black-headed gulls.

Under the shadow of Coniston Old Man I have watched powerboats noisily racing up and down Coniston Water, hauling water skiers behind them, but on Windermere beneath Cartmel Fell yachts with bright red sails make slow progress over a silvery pool. Common sandpipers come back each year from Africa, flicking along the lake shore or calling nervously from boulders lapped by the wavelets that end in a dying tinkle of sound. Windermere and Coniston are set among high and wooded hills and in their waters swim trout, pike, perch and char. Higher up the mountains the lakes and tarns with rocky shores have only trout and char. It is on these higher pools that one should listen for the honking calls of the greylag geese recently introduced to the area. Windermere supports a few duck in winter time while cormorants, gulls and divers can be seen fishing in the water, and I have records of teal and goldeneye as well. Mallard and red-breasted mergansers nest around some of the lakes, which are brightened with the flowers of water lilies and lilac-coloured water lobelia.

From Skelwith I often make my way up towards Blea Tarn and the Wrynose Pass. On the narrow road in spring every corner is full of foraging chaffinches, and one day a sheep joined them and put its two front hooves on the door of the car. Here buzzards soar overhead and cock wheatears throw themselves up in the air to begin their jaunty little ditties. Tall juniper scrub grows around Blea Tarn amongst heather and bracken, and the air echoes with the songs of meadow pipits and the wails and guttural complaints of curlews. The route swings down into Langdale and here I have scrambled up the scree which flows dangerously down from the Pike of Stickle to look for rejected and roughed out axeheads of volcanic tuff from the prehistoric stone axe factory. This was the centre for the making and distribution of the Cumbrian type of stone axe; tuff is a fine-grained greyish-green stone of the Borrowdale volcanic series. Ravens have circled above me as I made my search, and I have met with parties of swallows flying towards the Stake Pass. The air is often fresh and clear and on these mountains there is a great sense of freedom and space, although the Lake District itself is small and compact – like Snowdonia – and barely thirty miles across. I have looked down on fell walkers, bright in orange or red anoraks and cagoules, making their way ant-like to me on my high perch, along the path towards Dungeon Ghyll. Across the valley like sturdy welcoming friends lie Crinkle Crags, Bow Fell and Sca Fell and on the Pike of Stickle, like Wordsworth himself,

> I sit upon this old grey stone
> And dream my time away.

Yet for the naturalist there never seems enough time to explore the countryside and to try and learn more about its living creatures. The mountains themselves make good walking country if the normal rules of hill travelling are observed but the flora that greets the journeyer is not outstanding. Many of the high summits are covered with woolly fringe moss and I have also seen bilberry, cowberry, crowberry, viviparous fescue, mountain bedstraw and mountain sedge as well as heath, mosses and bents. Grazing by sheep has to some extent contributed to the impoverishment of the flora. Dorothy Wordsworth noted that on Sca Fell Pike 'not a blade of grass was to be seen', but wind, snow-lie and temperature all have their part to play as well. Corries and screes are often interesting floristically, and parsley fern, mountain fern and alpine lady's mantle grow on some of the block scree. Shrubby cinquefoil has a foothold on the Wastwater screes and red alpine catchfly on some of the Skiddaw slates; generally they grow out of the way of the grazing Herdwick sheep and the human collector as well. Many of the peaty upland pools have midges, water bugs and beetles and the mountain ringlet butterfly may reward the climber above a height of eighteen hundred feet or so. When I have wandered over the high Lakeland hills I have had ravens and crows, wrens, merlins, ring ouzels and golden plover as my companions. The dotterel may also have a tenuous hold and golden eagles began to nest again in 1969 after an absence of more than a century.

Unfortunately three young eagles have been shot and killed in the Lake District since the birds re-established themselves and numbers of buzzards are also being shot or poisoned.

There are scrubby woods growing up some of the Lakeland hills, where willow warblers and chaffinches sing all the spring wakening hours. In some of the steamy damp oakwoods I have found redstarts and wood warblers bringing to the hillsides some of the loveliest sounds of northern England, while the pied flycatcher is one of the most typical birds of the fell woods, along with coal tits. There may be red and roe deer in these high-level woodlands as well as red squirrels, foxes, badgers, stoats and weasels. Pine martens have been seen in Grizedale Forest, which is managed by the Forestry Commission and embraces both remnant broad-leaved woods and conifers planted on the higher ground. The fixing of several hundred nestboxes in Grizedale Forest has encouraged the breeding of pied flycatchers and great, blue and coal tits in areas where there were no natural sites for these hole-nesting species. Much has been written about the Lake District and there is much to be enjoyed within its borders – the glisten of the first frosting of snow on the high tops, still reflections of brown hills in tiny tarns, the rush of running, falling water, red-berried rowans and golden globeflowers in damp meadows, the green slates of Honister, the ruins of Hardknott Castle, men repairing stone walls with two sets of 'throughs' and cam tops, countless lambs among the daffodils, ring ouzels fluting on the summer fells, fieldfares at two thousand feet clustered on the field woodrush, Castlerigg stone circle and the dale-head farms of the National Trust.

Despite the ever-growing pressure from visitors, which demands both examination and some control, there are fortunately still some places which only two feet can reach! The landscapes of the Lake District have been much altered by man, who cut down the original forest cover, introduced grazing animals and quarried for stone, and planted woodlands of alien conifers. It is some of the old woodlands and wooded lake shores and the more accessible fells that are most likely to be degraded by man. Yet the diligent naturalist can still find some places which possess what Daniel Defore called 'a kind of inhospitable terror'. The Countryside Commission, the National Trust, the Forestry Commission, the local Naturalists' Trust, the Nature Conservancy Council, and Manchester Corporation are all involved in preserving the integrity of this part of England, and the late Sir Dudley Stamp observed that 'The whole National Park may be regarded as a nature reserve'.

The coastline of Cumbria also has several attractions for the naturalist. The rare dune cabbage grows here. There is a local nature reserve at Ravenglass and Drigg Dunes. The Roman port and stronghold of Ravenglass lies by the trident-shaped estuary of the River Esk. On the sand dunes across the water is a large black-headed gull colony and nesting groups of little, Arctic and common terns and sometimes Sandwich terns. Beyond and farther up the coast lie Sellafield and the tall symbols of Windscale and nuclear power. Two miles south of the busy seaport of Whitehaven the red sandstone cliffs of St Bees push out their 365-foot-high bluff into the Irish Sea, pointing to the northern tip of the Isle of Man. The red Permian sandstone cliffs are home for 2,500 guillemots and a few puffins, fulmars and cormorants, while black guillemots have made this their only breeding site in England. The RSPB has now set up a reserve at St Bees. Around the corner from St Bees' red cliffs lies the Solway Firth – the culmination of our journey from the sands of the Cheshire Dee. Forty square miles of the Firth have been declared an Area of Outstanding Natural Beauty and this includes Rockcliffe Marsh – a great winter haunt for wildfowl.

I came to know the Solway Firth from the air during the last war but I have made visits in autumn and winter, especially to the National Nature Reserve at Caerlaverock on the coast of Dumfries-shire, set between the shifting boundaries of the Rivers Nith and Lochar. It is a part of the world where wildfowling has been pursued for centuries, firstly for food and latterly for sport. At high water of an ordinary spring tide about one tenth of the reserve is left exposed as 'merse', and grazed by cattle, sheep and wild geese, and marked by pools and deep drainage channels. Sea meadow grass, sea aster and sea pink grow on the saltmarshes. Higher up are grassy swards with bents, sea milkwort, sea arrowgrass and chestnut sedge – the last as common at Caerlaverock as anywhere in Britain. Lapwing, oystercatcher, ringed plover, snipe, redshank, mallard and shelduck breed on the reserve and there are foxes, stoats, roe deer and hares. The reserve is also the most northerly point for the natterjack toad. Many kinds of wildfowl and waders can be seen along the Solway shore.

At nearby Powfoot early one October I counted more than a thousand golden plover gathered just below the sea wall; it was a fine, warm day and some of the birds were yodelling as if they had just arrived on their spring mountain breeding grounds. There were curlew, godwit, redshank, ringed plover and dunlin – in fact, after Morecambe Bay the Solway is the next most important estuary for waders in Britain. Pink-footed geese are regular between Caerlaverock and Gretna, greylag geese predominate in the western

Solway. The barnacle geese have undergone various vicissitudes, with peaks of around ten thousand on the Blackshaw Bank in the 1880s falling to only five hundred at the end of the Second World War.

Barnacle geese that nest in Greenland winter on the Western Isles of Scotland, while those that breed in Novaya Zemlya and district winter in the Low Countries. However, the barnacles wintering in the Solway represent the entire breeding population of Svalbard or Spitzbergen. I can remember early one October at Caerlaverock looking at some seven thousand barnacle geese newly arrived from Svalbard grazing furiously on the grass of two fields below the watchtower. They were in family parties moving steadily towards my point of concealment. There were two leucistic birds showing up white among those with normal colouring. All the time the air was full of their barks, yelps and humming sounds. Soon several score were feeding on both sides of a thorn hedge. Suddenly the geese moved away from the hedge like the fur of a cat parted by the wind, leaving an empty strip of grass either side of the field boundary. Robert Bridson and I waited to see the cause of this movement. Across the corner of the field stepped a female roe. She reached a tall wire fence and with a standing bound cleared it. The geese began to move back once more and it was with a rare sense of privilege that I watched probably the entire population from Svalbard spread in front of me.

I would like to end this chapter with an account of a visit I made near Christmas time to Caerlaverock. It was on a day of bitter frost and the rough grazings were iron hard and pitted with the frozen foot-prints of cattle. With the warden Robert Bridson I made my way along a track set between high hedges, which effectively screened us from the geese in the fields nearby. We could see redwings and fieldfares in the thorns in the hedgerows and a sparrowhawk flew powerfully ahead of us. Then we clambered into a small hide in the bank, provided by the Wildfowl Trust, which has leased Eastpark Farm. Only twenty yards away was a flock of barnacle geese – grey and black – spread out like a living carpet in front of us. I could listen to their monosyllabic barks – some high, some lower in pitch – while their conversations and altercations were punctuated by the deep lowing sounds from the cattle in the field. We then left the hide and made our way to the watchtower which overlooked Blackshaw Bank. From this high vantage point on the Saltcot Sandhills, where the natterjack toads were laid up for the winter, I could see curlew passing the window and farther away flocks of wigeon and mallard, dunlin and golden plover. The barnacles were a grey smudge on the green and brown field.

Wildfowling is a traditional sport on the Solway Firth; Wally Wright, who was born near Dumfries and started wildfowling when he was fifteen, finds the attraction of the pursuit compelling despite the uncompromising conditions of the Firth early or late and in freezing weather. 'It's the wildness of the geese and being out on your own', he said. To prevent too much disturbance to the wild geese a zoned system is used at Caerlaverock, with a limited number of permits for the central merse and no shooting on the merses on either side or on the sandy foreshore. Certainly relations between wildfowlers and conservationists here are very good and yearly counts are made on geese and duck numbers so that the situation is under constant review. There are similar reserves of the Nature Conservancy Council at Bridgewater Bay, the Dovey Estuary and Lindisfarne.

As the sun began to drop across the Solway I made my way to the shelter of some low gorse bushes to wait for the evening flight of the barnacles. A minute or so after half past four on that bitter December day a flock rose half a mile away and like a pack of barking dogs flew over my head; then flock after flock passed over, sharp against the sky in a wild chorus of notes – the sound of Christmas time in this wilderness of saltmarsh, sandbank, wind and tide.

11. The Backbone of England

Inland and roughly parallel to the north-west coast of England runs a chain of uplands, reaching from the Peak District of Derbyshire north to Cross Fell on the northern limit of the Pennines. The Peak itself is a term which generally applies to that region of the southern Pennines around the high plateaux of Kinderscout and Bleaklow. Within it lie the 542 square miles of the Peak District National Park – the first to be designated, in 1950. The Editors of Collins' New Naturalist series wrote that 'The situation of the Peak District in the heart of England as an island of varied hill land, often of spectacular scenic charm, almost surrounded by industrial lowland in whose cities and towns, often gloomy and grimy, live a quarter of Britain's population, made it a natural choice for the first of our National Parks'. It is a land of massive limestones, sandstones and 'grits', shales and volcanic rocks known in the region as 'toadstones'. Each shows a different resistance to the effect of erosion and so the Peak is a land of strange and delightful contrasts where farming, grazing, mineral extraction, forestry, water conservation, recreation and nature conservation are all reflected in the many faces of the landscape. The area is very popular especially during summer weekends, defying Daniel Defoe's reaction to this 'howling wilderness'; he certainly was no lover of mountains! On the southern boundary of the Park is the Domesday town which Dr Johnson called 'Ashbourne in the Peak'. The Peak denotes the region and does not refer to any special hill summits, which are often known as 'pikes' or 'tors'. Within the Peak are the towns of Bakewell and Buxton and the great houses of Chatsworth set above the River Derwent and Haddon Hall above the Wye.

The main structures of the hills of the Peak are the moors and rough pastureland of the hard, unyielding angular-grained Millstone Grit – the Dark Peak – and the green pastures and wooded dales of the Carboniferous Mountain Limestone – ancient sea-sediments 280 million years old heaved up at a later geological period – the White Peak. The former is marked by bare moors, rugged cloughs and bubbling becks – a scenario for *Wuthering Heights* – while the limestone scenery is softer, more gentle and sylvan, with slower flowing rivers. My first of many visits to the Peak was in November. I started in Dovedale, a gorge-like valley with wooded slopes of clinging ash trees, wych elms, elders, hazels, rock whitebeams, and yew on the craggy outcrops. Bastions, buttresses and pinnacles of rock soared up above the little dingles and the river. The cone of Thorpe Cloud, a typical reef knoll less soluble than the surrounding limestone which guards the entrance to Dovedale, rose up nearly a thousand feet. As I walked along the riverbank a kestrel, occupying one unchanging piece of space, hovered on an icy up-current of air. Grey wagtails called 'zit-zit' and a kingfisher flashed cobalt-blue like an electric spark upstream. A dipper bobbed and jerked on a stepping stone and its cheerful little conversation of a song added charm to the cold autumn day. The limestone rivers are often fed from underground and tend to be somewhat warmer than those of the grit; the Wye is 4° F higher than the Derwent at the same height. The limestone rivers of the Peak contain three times the amount of suitable food for dippers that the gritstone rivers do. It is the shallowness of the water which seems to attract the dippers, however, and a pair per mile seems to be a fair average.

In summer the slopes of Dovedale rejoice in a delightful limestone flora, and for me the rewards are spring cinquefoil, Nottingham catchfly, herb paris and greater burnet saxifrage. Water buttercups, fool's watercress and pondweeds flourish in the stream, and along the bank I have seen yellow iris, loosestrife, meadowsweet and many kinds of sedge. In the woods there are tits, willow warblers, tree pipits and chaffinches. In Lathkill Dale, just over the hump of Arbor Low from Dovedale, are relict ashwoods in a National Nature Reserve, where there are also redstarts, wood warblers and hawfinches, and the ground flora includes fine-leaved sandwort and Jacob's ladder. Monk's Dale has willow tits and a very fine range of limestone plants. It is the staggering variety of these flowers which makes the Peak so attractive to a naturalist like myself – mountain pansy, globeflower, common lady's mantle, lily-of-the-valley, giant bellflower, dropwort, melancholy thistle and bloody cranesbill – all gems of the limestone scene and here sometimes in profusion. For the traveller who wishes to enjoy these flowers June and July are the best months, when the lapwings wheel and tumble, the curlews rise up in their display flights and the wheatears build in the stone walls.

A tributary of the River Dove is the Manifold – a river indeed of many folds – where, as in the Dove, trout and grayling swim in Isaak Walton's 'clear waters of Derbyshire'. The Manifold rises near Axe Edge and pursues a course of great beauty through a green valley. Between Thor's Cave and Wetton Mill it disappears into 'Swallows' and reappears some four miles farther on. In winter I have seen the hawthorns here filled with fieldfares, while rooks and jackdaws assembled in the fields near the river. From the valleys of the Manifold and the Dove I have moved up to Arbor Low – the great stone circle of some forty limestone blocks all lying upon the ground and pointing towards the centre; the monument stands on a hilltop more than twelve hundred feet high. On one visit I found the circle guarded by an enormous shaggy bull. It is one of the largest Mesolithic or early Bronze Age stone circles in England but it lacks the drama of Avebury and Stonehenge.

After walking round the circle I made for Castleton, which lies at the edge of the Hope Valley and in the centre of the most hilly part of the Peak. It nestles between the grit hump of Lose Hill and the stony furrowed valley of Cave Dale. In the past lead was regularly mined near Castleton; it was found as galena in veins filling vertical cracks, known as 'rakes', which appeared when molten material flowed at the same time as the Carboniferous limestone was being formed. Besides lead there was often other material such as barytes, calcite and fluorspar. Lead mining reached its height in the late eighteenth century but declined after about 1850. A special and rather beautiful variety of fluorspar with a purple or amethyst colour is still mined to make ornaments; it comes from the Blue John Mine near Castleton and opposite 'The Shivering Mountain' – Mam Tor – where soft shales below hard sandstones and grits are washed and weathered away so that the upper parts are undermined and fall down the face of the mountain. Tables, vases and candelabra were made and exported especially to France, where some of the fluorspar was used in massive ormolu work; here it was known as 'bleu-jaune' – perhaps the origin of the popular name. Arthur Ollerenshaw, the proprietor of the Blue John Mine, created a silver bracelet with ten matching pieces of Blue John for my silver wedding present to my wife. As you gaze into the translucent white and amethystine crystals, it is like peering into fractured ice stained with toothed bands or hazy streaks of purple offering a vista of miniature mountain ranges and valleys leading away into the distance. Above Castleton lies Peveril Castle, built according to Sir Walter Scott 'on principles upon which an eagle selects her eyrie' but I have found only chattering starlings. In the villages the grey limestone walls of the cottages hold roofs and lintels of the Millstone grit.

The countryside of the 'grit' is a bleaker one altogether, with higher, more rugged scenery. Here grazing by sheep and the protection of the grouse have helped to influence the landscape; the former has led to barren moors where skylarks and meadow pipits may be the commonest birds and the latter, with its cult of a gamebird, to rotational burning of heather and persecution of the birds of prey. The moors may be clothed in heather or carpeted with cotton grass, where the cotton sedge moth flourishes, and peat hags. I have looked down on the heather-covered plateaux of Kinderscout and Combs Moss, where the rolling heathery slopes carry bilberry and crowberry, and rock-strewn gullies or cloughs fret the landscape. On the lower grit slopes curlew, lapwings and snipe breed, while on the higher ground golden plover are widespread. Once the attractive golden plovers could be found nesting at just over 800 feet but now you have to walk up to 1,150 feet or more to find their territories. Like the red grouse – its regular companion – the 'goldie' is very much a bird of the moors, heather and blanket bogs gay with cotton grass. Near the soggy peat hags of Bleaklow there are those tiny trilling waders, the dunlin, as well. Heather moor is rather typical of the western and northern slopes of Kinderscout and Bleaklow, while cotton grass, whose nodding white heads transform a moor into a terrain of snow, is dominant elsewhere on Bleaklow, Combs Moss, above Alport and at the head of the Goyt Valley, which is closed to traffic in summer except for a limited number of minibuses along a restricted section of road. Bilberry moor can often be seen alongside heather, and mat grass is common. The lichen flora on Kinderscout is rather poor and this has been attributed to the loss of peat cover and aerial pollution.

Some of the upland streams hold dippers, wagtails and common sandpipers, while ring ouzels may take up their territories and sing amongst the block scree of the rugged cloughs and clefts on the hills. On some of the rocky, heathery slopes there are perhaps half a dozen pairs of merlins but the numbers of this game little bird of prey have been going down in recent years. It is a rare and rewarding sight to glimpse a merlin in the Peak hunting boldly and with élan over the moors. One of the more interesting birds of the Peak District is the twite or mountain linnet which is widely established along the ridge of western hills from Glossop to Axe Edge, breeding on both heathery grouse moors and wet and misty cotton-grass mosses. The

165

birds rise up in front of you rather like linnets but I find that their calls are more nasal and metallic and their songs are rather jangling and resonant. On some of the gritstone slopes there are oak and ash woods where the birdwatcher will find pied flycatchers, redstarts and wood warblers – I have seen all three species in one small wood – and there are nightjars and some black grouse along the edges of these woodlands and Forestry Commission plantations; the State Forests attract many goldcrests and coal tits, redpolls, sparrow-hawks and sometimes long-eared owls. Badgers are not common but I have seen foxes running strongly over the fields and slopes, weasels tripping along the tops of stone walls and I have glimpsed the introduced mountain hares and seen their corpses on the roads. I have not seen the red-necked wallabies living wild and free on the Derbyshire moors near the Staffordshire border.

North-east of Castleton are the reservoirs of Derwent Dale, form-ing a chain of lakes some seven miles long. Here is a Lake District in miniature with the dark greens of spruce and pine lit up in autumn by the golden larches and the bracken in decline. There are often no birds on these waters on the Millstone Grit and the small numbers that do occur may include moorhen and mallard, a few teal and tufted duck and an occasional diver. When I visit these reservoirs in the Derwent Valley I usually pause for sad and silent contemplation of the little memorial stone to the sheepdog which one winter gallantly guarded her dead master for fifteen weeks – from 12th December to 27th March. It is a spot where magpies rattle and linnets wheeze. One of my regular journeys takes me north-west along the Roman road to Glossop – the Snake Road – where in summer ring ouzels can be heard singing. Here the moorland around Jacob's Ladder is the home of the red grouse, which cackle and laugh at the lonely traveller on the hill, comforting him in kind weather and sending him back home in bad!

There are many delightful parts of the Peak – Water-cum-Jolly Dale is as attractive with its sylvan scenery and rare limestone grass flowers as its name suggests; Chee Dale has its massive cliffs and abundant bryophytes; Hob's House is rich in Mountain limestone corals; Cressbrook Dale also has many flowers and rare plants; the Goyt Valley has ancient oakwoods and rather special moths; Moss Carr rejoices in marsh cinquefoil, bogbean and orchids; Alport Castle falls slowly away from the mountainside as a shifting serrated gritstone tower; Tideswell boasts its church – the Cathedral of the Peak – and its well-dressing ceremony; Axe Edge remains more or less as it was when W. H. Hudson came to study its birds; Buxton lies outside the Park but rests on its reputation for water and air; at

Stanage Edge lie incomplete millstones of that same grit which was used to build Chatsworth and the Crescent at Buxton – a whole host of wonders for the inquiring, wandering naturalist.

The Pennines, of course, stretch to the north as far as Cross Fell. They are formed from the Carboniferous or Mountain Limestone, the Yoredale rocks of shale, sandstone and impure limestone and over the summits the Millstone Grit. The limestone arose from a clear sea full of corals, the Yoredale series from a river, and the grit from deposits of sand and mud over an area of more than 25,000 square miles. From Kinderscout in the Peak Fell the grit produces the 'Edges' or moorland escarpments which are so typical and caps such hills as Ingleborough, Great Shunner and Cross Fell. As the Carboniferous Period drew towards its close a great sheet of plutonic dolerite or basalt – a mighty lava flow – forced its way horizontally between the bedding places as a 'sill', which being resistant in a series of softer sedimentary layers, gave rise to a sharp scarp; this Whin Sill runs under northern England from the Farne Islands to Teesdale and Cross Fell and has enhanced the landscape, with its bluffs, which carry Hadrian's Wall for many miles, its steps which have made the waterfalls of Caldron Snout and High Force, and its columns which fringe Roman Fell and the wonder of High Cup Nick near Appleby. There are beds of New Red Sandstone in the valley of the Eden but the former coalfields have been eroded away and survive only on either side of the Pennines. Mountain movements and folding produced further changes and the later ice sheets scoured the hills, widened the valleys and left behind countless moraines and piles of glacial drift. The Pennines have been called 'the remains of a plateau in which rivers have carved their valleys'. There are long stretches of moorland, valleys that dip away to the horizon, cloudscapes of never-ending variety.

Conditions up here may be hard; the land is not fertile and the growing season is short. The wind may be severe and exposure high but there are days when the sun shines after the mist and rain and the soft susurration of the breeze in the heather or the mountain grasses soothes the spirit and refreshes the mind. In winter the snow lies deep over the black heather and silvery grass, and only a few species of animal are able to stay on the hills all the year. In spring a pale green spreads over the plateau while in August the heather blossoms in a riot of purple finery. In summer waders, pipits and wagtails return to the peat bogs and mosses, where cotton grass, bog myrtle and bog asphodel flourish, or to the swards of mat grass and fescue.

From Edale and Kinderscout in the Peak a continuous footpath known as the Pennine Way runs to the Scottish Borders. Completed in 1965 this track meanders along the backbone of England, providing wide views and an intimate introduction to the hills first known in the fourteenth century as *Alpes Penina*. It skirts Crowden and the impressive valley of Longdendale, with Manchester's reservoirs and Saddleworth Moor, climbing up Black Hill and leading on to Stand Edge. From the hilly track it is soon possible to look down on industrial Lancashire, while ahead unfold vistas of heather moor and cotton grass where skylarks and meadow pipits bring some cheer to the wilderness. From Longdendale northwards it is worthwhile keeping one's eyes open and ears alert for twites. Beyond the M62 Motorway lies Blackstone Edge, which the writer and traveller Celia Fiennes described as 'a dismal high precipice'. Farther on is a complex of four reservoirs which, like many in the Pennines, are somewhat oligotrophic – clear, high up but rather low in nutrients – and a few mallard are often the only birds present. Near Hebden Bridge the hillsides become chequered with small fields and drystone walls where wrens and a few wheatears nest in the summer. Around Hebden and Heptonstall the narrow Pennine valleys with their stone cottages reflect the coming of the Industrial Revolution, which also brought the mills; both domestic and manufacturing architecture reveal the use of the solid Millstone Grit in walls, mullions and flagstone roofs.

North-west of Hebden Bridge is a favourite walk along Hardcastle Crags, where on a spring morning one can hear coal tits and goldcrests wheezing in the pines; this region has been called 'little Switzerland'. This is a high land of packhorse tracks – the home of the Brontë sisters. Here the Pennine Way passes through Cowling and Lothersdale, where the laburnum blossoms yellow in the gardens, and up to Pinhaw Beacon. From this 1273-foot eminence one can have views across to Bowland, while Ingleborough and Pen-y-Ghent lie ahead like eroded sand castles on a giant beach, and to the east rises the limestone plateau of Malham and Great Whernside. Skipton is now only four miles away, a gateway to the Yorkshire Dales, both milltown and market, with two great focal points – its Norman fortress and restored Church; once Sceptun – the Sheep Town of the Angles – Skipton still has its stock auctions.

Here lies the Aire Gap where that river has cut through the Pennines – an important route ever since traders of the early Bronze Age made their way from Ireland across to Scandinavia. It is still important today for the ornithologist, since many lesser black-backed gulls use the Gap in spring to pass through the Pennines, while in July large numbers of shelduck from Morecambe Bay fly east through the Gap to moult in the eastern parts of the North Sea.

Thousands of shelduck fly through the Gap and may be using the conspicuous and unusually shaped mass of Ingleborough as a landmark.

The backbone of England, once dark acid bogland of the grit country, now begins to change into the varied flower-bedecked and softer country of the limestone of the Craven district of Yorkshire. The landscape is still an upland one and the limestone reaches up towards the heights of Pen-y-Ghent, Ingleborough and Whernside, whose flat heads rear up above the green calcareous grasslands. The summits are cut off from each other by river valleys – Wharfedale, Littondale, Ribblesdale and others. This is a land of contrasting colours like the Peak District. White gleaming scars and cliffs of limestone, topped with pale green grass and limestone mires, lead up gently towards the peaks of gritstone or softer shale which weather into dark screes and produce thin soils and a peaty eroded summit plateau. Streams rise and plunge into the limestone, while below the hills the more lowland areas show good productive farms and small woodlands very different from the scenery of the treeless uplands. For the walker interested in flowers Craven is not to be missed. Magnificent and often awe-inspiring cliffs are the homes of some of our most exciting plants and here I have enjoyed some of my most rewarding days in the field.

Near the village of Malham in Yorkshire is a ridge of limestone formed as the result of a fault in the rocks millions of years ago. In several places the limestone has been cut back by water – at Gordale Scar, Yorkshire's Grand Canyon, with its deep dramatic rift in the hills and its tumbling beck, and at Malham Cove, once perhaps the highest waterfall in Europe but now dry. From the bottom of the Cove a beck now gushes out. Malham Cove rises as a moss-stained white cliff 240 feet above the stream – a magnificent natural amphitheatre at the head of the Dale and one of the finest sights in Britain. Big screes on the valley sides hold or lead down to open woods of ash, hazel and hawthorn. I have botanized along the slopes and marvelled at the blue sheets of Jacob's ladder – carpets of tall clustered flowers growing on the scree just as John Ray described them in 1677 'about Malham Cove, a place so remarkable that it is esteemed one of the wonders of Craven'. I have also found alpine pennycress, bloody cranesbill and whitlow grass as I crossed the rocks in the beck and explored the banks; above and to the east are traces of lynchets or farming terraces. It is a wonderful experience to approach the great cliff and look upwards. I have watched house martins building their hemispheric cups of mud under horizontal ledges on the face and one year a kestrel had a nest there.

Above the Cove is a flat plateau of large limestone blocks separated by cracks and crevices often several feet deep and known as 'grykes', but this region has been rather heavily grazed. The slopes of Gordale are covered with fescue and blue moor grass, favoured by the large heath butterfly which is a feature of Craven, and interrupted by limestone outcrops and screes. Yew and rock whitebeam cling to the canyon's sides and there are remnant ash woods on the steepest sidings, and bents and matgrass with its fox or antler moths on the summit plateaux. To the east in Wharfedale I know of places where lily-of-the-valley, solomon's seal and fly orchid grow, and in the fragmented woods that form their habitat there are also jays, willow tits, great spotted woodpeckers and woodcock. The once declared home of the rare lady's slipper orchid, this is a dale where the patient searcher may find bird's eye primrose – often common – and rarities like the bitter milkwort, angular solomon's seal and dark red helleborine. In spring Littondale is awash with snowdrops, and on the high limestone to the west mountain avens has a toehold.

There are wide uplands from Malham to Littondale. To the south of the Craven Fault lie the Millstone Grit and Yoredale shales, marked by heather and bracken and swampy regions of bent and rush, and in contrast to the pale green limestone grasslands. One can almost trace the fault line by the walls of limestone or grit. Malham Tarn lies at twelve hundred feet on the high ground – a natural lake resting on the Silurian slates higher than the limestone to the south. In the Tarn I have caught loach, bullheads and crayfish and in the countryside around listened to and filmed the curlews bubbling in the mountain air, meadow pipits and wheatears rising up in their song flights, golden plover whistling on the stony Tarn edge and great crested grebes nesting and hatching their young in the fenny fringe where bottle sedge, bog bean, marsh cinquefoil and angelica form a watery jungle in which sallows develop their grey canopies. There are mallard on the Tarn and small numbers of coot and moorhens and occasional tufted ducks. Goosanders are rare visitors but I once watched a dunlin on the nearby moorland dragging a wing on the ground in front of me, trying with its feigning of injury to draw me away from the hard-set eggs or young chicks. High up on the moor there is a 'lek', or display ground, for black grouse.

Tarn House overlooks the water, and guests at what is now one of the Centres of the Field Studies Council once included John Stuart Mill, John Ruskin, Thomas Hughes and Charles Darwin. Charles Kingsley's Lowthwaite Crag and Vendale in *The Water Babies* recall his visits to Malham and Littondale respectively. The skyline beyond the Tarn is dominated by Fountains Fell and Pen-y-Ghent – grit caps on limestone. There are unrivalled views as you walk towards the summit of Pen-y-Ghent across blanket bog on glacial drift, with carpets of cotton grass – so poor in invertebrate life – and heath rush, cloud berry and both purple and yellow mountain saxifrage, over limestone grassland below the cliff summit, and then across bents and matgrass near the top. Pen-y-Ghent lies like a resting lion across the backbone of England; often deep in winter snow it gives way to a succession of flowers in summer – violet and primrose, wood sorrel and windflower, bird's eye primrose and baneberry, rock rose and cranesbill – in a land where sheep graze, curlews 'whaup' and black-headed gulls from the plains below scream and forage on their narrow white wings. This is the very heart of Craven and here Pen-y-Ghent – sharp-snouted and sha-ttered, Whernside with its flat cap of gritstone and Ingleborough, less grand and described by W. R. Mitchell as 'a green-fleshed whale', raise their isolated summits. I have seen the last of these summits streaked vertically and horizontally by bands of snow; the approach to it takes you across a flat stony plateau and up through mixed moorland to mosses of dancing, bobbing cotton grass. This is the home of the rare Yorkshire sandwort, and on the long Ing-leborough Scar is one of the great treasures of the Yorkshire Dales National Park: this is Colt Park Wood, an ash wood at eleven hundred feet growing on high limestone pavement. Besides the ash trees in this strange open jungle of scrub there are bird cherries, rowans, guelder rose and thorn growing partly in the accumulated humus and partly in the grykes. In these cracks flourish large amounts of male fern, ramsons, lily-of-the-valley, baneberry, giant bellflower and there are clumps of large-flowered pansy, yellow star-of-Bethlehem at its highest situation in the British Isles and in the wetter sites meadowsweet as well.

Not far away is Ling Gill, which like Colt Park is a National Nature Reserve; it has impressive wooded banks and boulder-strewn scenery – the home of the melancholy thistle and other rare plants growing under the ash and birch trees. From here the spine of Pennine uplands stretches to Cross Fell, often wet and in cloud and scoured by wind and frost, bronzy-green under the bog moss, black under heather or purple in its flowering, criss-crossed with erosion channels and open to the sky. The wading birds that live up here have developed wild far-carrying calls to suit the wide open spaces and their fluted or trilling sounds are an essential ingredient of the upland scene. In their greys, browns and blacks insects and other small animals live unnoticed in their high habitats. The slow decay

171

of moorland plants means only a small population of tiny animals. There are several species of moth, plant bug, bee, fly and spider which the walker may come across, and some of the upland pools have water beetles and dragonflies. There are great expanses of fell, deep valleys, waterfalls and caves. Upper Wharfedale, once known as Langstrothdale, combines green hillsides, narrow woodland groves and limestone terraces set above a beck that as one follows its course dashes from cascade into shallow pool and through narrow ravines, between birches and ash trees and around rocky boulders, unfolding a continuous strip of changing scene and wonder.

Beyond lies the broad channel of Wensleydale, the widest and softest of the Yorkshire Dales; it is the course of the River Ure – more anciently the Yore – which runs from the Moorcock Inn east to Leyburn. Charles Kingsley observed that 'the richest spot in all England is this beautiful oasis in the mountains'. It enjoys many tributary valleys – Coverdale rich in medieval remains, Waldendale and Bishopdale with their secret farmsteads and Cotterdale with its Primitive chapel. Here life still goes on in an enviable seclusion and peace. Wensleydale is also rich in castles. Bolton was in part quarried from nearby Apedale Greets and here Mary Queen of Scots was held 'in honourable custody' after her defeat at Langside in May 1568; today the fortress stands square and high above Wensleydale and the jackdaws and pied wagtails disport themselves around the ruins of the Great Hall and the belfries on the south-west tower. There is also Middleham Castle, home of the Nevilles and Richard, Duke of Gloucester, later Richard III, set amongst Georgian houses and racing stables.

The Ure rises on the high fells in the shadow of Great Shunner and then wanders down into Wensleydale. The villages can be found along its banks or raised up on the valley sides, set in green spaciousness beneath the shadow of the dale slopes formed from the Yoredale shales, sands and limestones. Gills and falls streak the valley walls, strips of wood climb skywards up the bank, and farms and stone barriers speckle and streak the green slopes with particles of grey. In autumn colours West Witton is one of the showplaces of Wensleydale and the village from which the Dale takes its name – Wensley – lies on the northern slopes and looks out on a trim, well-tended green. Bainbridge, where I have stayed several times, also overlooks a green where contented fat cattle graze, daffodils spread a golden carpet under the trees and pied wagtails sit on the old village stocks when the children are at school. In spring Aysgarth is bright with violets and primroses and the burgeoning ash trees, while grass of Parnassus appears in the swamps.

On the hills grow scabious and harebells, foxglove and bellflower and in late summer ragwort provides crowns of yellow gold along the river. Marjoram grows at Jervaulx Abbey – home of the Cistercian monks – and the grounds are rich in wild flowers; more than a hundred different kinds were found in a season. The Dale, of course, is justly famous for its cheese, first made by the monks of Jervaulx, and its special flavour depends largely on two aspects of Wensleydale which already concern the naturalist – the rocks and soil, which influence the character of the pastures. Under the guidance of that fine Dalesman Kit Calvert I once took part in a television film about the cheese industry in the market town of Hawes – 't'Hass' as older people call it – perhaps derived from the word used in Anglo-Saxon to denote a pass through the mountains. The upper part of Wensleydale is marked by spoil heaps on the slopes, which speak silently of the past quarrying of the Yoredale sedimentary rocks for walls, buildings and roofs.

The River Ure may reveal little of itself to those who keep to the roads but in less accessible places the swift-flowing waters create dramatic cataracts like those at Aysgarth – the Norse 'clearing among the oaks' – where, at three points, the river pours over rock steps, and which Thomas Pennant in 1772 called 'most commonly picturesque'. Besides grass of Parnassus, which I find a truly romantic flower, there are dippers and grey wagtails bouncing and bobbing on the water-splashed boulders. At Hardraw a river falls ninety-eight feet over a limestone step – an unbroken cascade that in prolonged frost turns into a giant icicle. Above Bainbridge lies Semerwater – 'Deep asleep, deep asleep' – from which the River Bain flows noisily down some three miles across lush farmland where the swallows and house martins hawk for insects, past streams and ditches full of meadowsweet and wagtails, and verges rich in flowers and dancing, winking meadow browns and small heaths until it reaches the Ure. Semerwater is placid and shallow, the site of a former lake settlement, and its nutrient rich waters foster a growth of plankton that supports a varied community of aquatic animals. The abundant mayflies help to feed the bream and introduced perch that swim below the calm surface among the foraging crayfish. In summer this is the haunt of mallard and reed buntings, great crested grebes, dippers, grey wagtails and common sandpipers, while on the hills around I have seen and heard buzzards as they mew and soar on broad-fingered pinions, lapwings as they call 'pee-wit', curlews as they float down in spring display and the occasional raven cronking as he made his way over a lake turned by a freshening wind into creamy foam as white as a vat of Wensleydale cheese. In winter there are sometimes wild swans – whoopers fairly often and Bewick's less so – which drop in to explore Semerwater, and it may be possible to see wigeon, teal and goldeneye and perhaps grebes and divers too. It has been said that on calm evenings you can see the roofs of a submerged village under the water and hear the muffled sound of church bells, while Semerwater stays 'deep asleep till doom'. All the way along the River Ure the 'little dales' branch off and climb towards the fell country, and each is a world unique in topography and sometimes in natural history.

One of my favourite routes out of Wensleydale begins near Hardraw, crossing the tops to a spot between Thwaite and Muker in Swaledale. As I travel up the road I can look across to Great Shunner Fell, where the Pennine Way rises to its summit at 2,340 feet and whose slopes are the site of ancient cairns. Behind lies Upper Wensleydale and views across the solid masses of Great Whernside, Ingleborough and Pen-y-Ghent. I can also see the long dome of Pen Hill – a beacon in Napoleonic times. My route takes me through the Buttertubs Pass and I always stop to inspect the 'buttertubs' themselves – a series of vertical fluted shafts that plunge down into the limestone. Ivy and ferns clothe the steep walls, which drop sixty feet into the darkness. A warden of the Dales National Park once heard a rumble and witnessed the creation of a new 'buttertub'. The Pass is an impressive lonely place; the fells provide ideal ground for the Swaledale breed of sheep – black-faced and grey-nosed with laid-back horns. The nearby ravine in early summer echoes with the wild flutings of ring ouzels, descendants perhaps of birds familiar to those pioneers of wildlife photography, Richard and Cherry Kearton, who lived in the neighbourhood. Swaledale below is a compact dale marked with limestone scars and the dramatic aftermath of lead mining. It is narrower and its slopes are steeper than those in Wensleydale. Thwaite, Muker and Marske shelter in their daffodil-bedecked hollows and Richmond Castle dominates the Georgian and Victorian stone buildings and cobbled market place of the town that guards the entrance to Swaledale. In parts its green and pastoral landscape is full of familiar bird song – from blackbirds, thrushes, robins, wrens and hedgesparrows – while long grey walls climb the steep slopes to the surrounding fells and moors.

During the last war when I was on leave from the RAF I used to visit my home in Appleby in Westmorland. I travelled on the railway from Skipton and Settle that puffed its way through Mallerstang to the Eden Valley and Kirkby Stephen. Here I enjoyed my first experience of mountains, and wrote of 'ice-smoothed blocks of stone, eroded and bare near their misty and snowy summits and streaked with lines of greyish scree and drifts of dead brown bracken'. To the west rose the rolling Howgills and Wild Boar Fell, whose lip of Millstone Grit lowered over the valley. To the east were Mallerstang Edge, Hugh Seat and High Seat and Pendragon Castle, said to have been built by the father of Arthur, the Once and Future King, and certainly constructed to guard the Pass against the marauding Scots. The deep furrow of Mallerstang nurtures the infant Eden and the isolated trees and farmsteads that stand square and unyielding behind their stone walls. High above the sounds of curlew and lapwing are countless skylarks and pipits on the high fells, while ravens, dunlin and golden plover breed on the plateau flats and rocky outcrops. Sometimes there are snow buntings in the winter. Although there is quite a lot of suitable ground for nesting dotterels I have not seen them in the Pennines here but there is always a possibility that one might come across a 'trip' – a small party of birds going south in the autumn. There are many mosses and bogs in the region of Birkdale Common, Stonesdale Moor and Arkengarthdale Moor which might attract the migrants.

North of Brough and Stainmore the Pennines become some of the most remote and wild uplands in England. Appleby, which was my home, and the then county town of Westmorland, lay under the shadow of Lady Anne Clifford's castle. Wordsworth admired this fortress on a bend in the River Eden, writing of

> She that keepeth watch and ward
> Her statelier Eden's cause to guard.

I lived at Castle Bank in the 1940s and returned in 1978 to stay in the castle, which is now a registered centre of the Endangered Species Survival Trust. The main street of Appleby, Boroughgate, is graced with pleasant Georgian houses and leads steeply up towards the great Norman keep of the castle. At its lower end stand the ancient Moot Hall and the church and cloisters of St Lawrence, rebuilt in the late twelfth century after being burned by the Scots. Set at the top, facing the road, is a pillar bearing the legend 'Retain Your Loyalty, Preserve Your Rights'. Salmon would jump the weir at Bongate and grey wagtails called along the Eden's banks. In summer swifts used to scythe the air and scream in wild abandon around the house roofs. The road verges were deeply scented with orchids, growing in profusion in the grass, and on the higher ground were lady's mantle and bird's eye primrose. I explored the wooded Vale of Tempe and walked along Bandley Beck to Burrells to look at the classic exposures of the lower Brockram rocks. Flakebridge Woods were full of bird songs and roedeer, and black-headed gulls screamed in torrents of white wings on Brackenber Moor. The Vale of the Eden lies on red sandstone and with its river, marshes, grassland, arable, rough grazing and stone walls supports a wide range of different bird species. Skylarks were the commonest birds, lapwings bred on the cultivated soil, and redshank, curlew, snipe and meadow pipit seemed to prefer the rough damp pastures. There were also whinchats, yellow and pied wagtails, reed buntings and occasional corn buntings, partridges, mallard and a range of woodland and hedgerow birds; however, these birds that favour 'steppe' conditions are much commoner than on farms in lowland England.

From my home in Appleby I explored the hills on foot from Lune Moor to Cross Fell. In those days I could travel anywhere but now red flags warn the hill walker of Danger Areas and firing ranges. I used to make my way up Hilton Beck between the limestone bluffs of Murton Fell to the left and Hilton Fell to the right. Three pairs of dippers held territories along the beck's tumbling bubbling waters but in 1978 I could find only one pair. As I used to walk along the beck, a dipper would fly ahead of me until he reached the end of his territory; then he would turn and wheel back in the air to pass over my head. Sometimes I would see a bird like a plump brown and white wren bobbing on a wave-splashed rock or swimming in the beck, turning over stones for the larvae of stone and caddis flies. In Westmorland the dippers' nests are often situated on rock faces, under bridges, on walls and even under waterfalls. Curlew and lapwing are constant companions and grey wagtails flick their long tails from the stone walls or wet rocks. Buzzards circle over Swindale Edge and once I regularly saw a peregrine; today a golden eagle from Lakeland may make a short visit.

I always used to stop at the disused lead and barytes mine where thirty years ago I could fill my pockets with heavy lumps of galena and white and amber crystals of calcite and fluorspar. In late winter I explored the fells above Swarth Beck, walking across frozen peat hags and snow bridges across deep ravines concealing streams far below the surface. It was a wild and stirring region of cold desolation. There were meadow pipits and red grouse up to two thousand feet, snipe up to fifteen hundred feet and occasional twite and snow buntings up to 2,200 feet; at this time of the year even the hardy sheep rarely ventured above eighteen hundred feet. In spring late snow and mist might delay the return of the hill birds and for the fell walker it was a forbidding terrain. My attempts to reach Mickle Fell, Cronkley Fell, Caldron Snout and Teesdale were often difficult and laborious and not always successful. Detours were often imperative but the expanse and loneliness of these regions of black and white landscape in which I would see no other living soul used to give me a tremendous sense of uplift and wellbeing.

In summer conditions are easier and the area around the Cow Green Reservoir, built after a long battle to save the Teesdale ecosystem, is well worth a visit. Here on the tops of hills like Mickle Fell is a granular crystalline limestone which looks like coarse granulated sugar and is known as 'sugar limestone'. Widdybank, Mickle and the other fells of this kind have been described as 'the Westminster Abbey of botany' since on them grow some of the most unusual and attractive plants in the British Isles. This herb-rich grassland was once exploited for cattle and even domestic geese. The sugar limestone is also well drained and you can easily spot it by the well-grazed turf set in a frame of dark bogland. Teesdale is a *locus classicus* for shrubby cinquefoil, bird's eye primrose and spring gentian but on the higher ground are two specialities – Teesdale violet and bog sandwort, besides hoary whitlow grass, mountain avens, alpine bartsia and several interesting rushes and sedges.

The Upper Teesdale National Nature Reserve has over eight thousand acres of sheepwalk and grouse moor but some of the land over which I once walked was drowned by the Tees Valley and Cleveland Water Board to form the Cow Green Reservoir. Here was a unique Ice Age community of plants. Many were removed to new sites but no one knows what the long term effects of a new and large area of water at sixteen hundred feet might have on so small and specialized a habitat. This high ground is still a home for ring ouzels, that nest in rocky clefts and even old mine shafts, curlew, golden plover, lapwing and snipe, although the waders other than the golden plover often favour the lower swampy ground. For me the reservoir and its presence have taken away some of the remoteness and sense of isolation. Caldron Snout with its white waters tumbling dramatically down over shattered angular steps in the Whin Sill still rewards a journey over the peat hags.

Another long but fascinating walk used to take me from Hilton Beck right across by Caldron Snout to High Force, where a path through the pinewoods with their red squirrels, coal tits and goldcrests would bring you to the fall itself; here the River Tees either dribbled over a great cliff in the Whin Sill in dry weather or, after a thaw or heavy rain, foamed and raged in a mighty awe-inspiring torrent as it plunged into a dark boulder-strewn pool. Above High Force is a good growth of juniper scrub where the shrubs stand like parasols or churchyard yews up to sixteen feet in height. One April I had eleven song thrush contacts in the juniper with eight for blackbirds, five for robins, hedgesparrows and meadow pipits, two each for greenfinch and redpoll and single ones for chaffinch, whinchat, woodpigeon, snipe and red grouse.

To the north-west of Appleby is High Cup Nick – an extraordinary horseshoe-shaped valley cut into the Pennines and with a high parapet of columnar cliffs of the Whin Sill. High Cup Gill meanders down the ice-cut valley towards the pastures, meadows and lines of trees – mostly ash and sycamore – in Eden Vale. The Pennine Way skirts the lip of High Cup Nick and from it on a clear day there are fine views of the Lakes hills from Skiddaw to Coniston Old Man and, of course, nearer at hand down into the Nick itself. At the foot are several small conical hills – Dufton Pike, Murton Pike and Knock Pike.

In the villages that cluster around these foothills there are countless stories of the dreaded Helm Wind, which sometimes comes down from Cross Fell with such force that trees and carts may be blown over and horses have been carried over precipices. The wind is marked by a long bank of cloud known as the 'helm' lying along Cross Fell, while a few miles to the south-west is another cloud bank called the 'burr' or 'bar'. The wind is believed to be caused by cold air from the east coming down the hillside, being warmed up at lower levels and starting to rise once more; in effect, a standing wave is set up. Perhaps the Helm Wind gave rise to Cross Fell's older name of Fiends' Fell since it rolls down like thunder; it is very much a feature of the northern Pennines in late spring and autumn.

My recent visit in 1978 took me back into the fell country in the month of April. Buzzards and kestrels still quartered the hillsides, wheatears flashed white rumps from rock to rock, meadow pipits went up in aerial song, dippers bounced on the becks, curlews bubbled and lapwings fell about in the sky over the rough grassland. A ring ouzel still sang up Hilton Beck. On great Dun Fell with its masts and radar station my wife and I walked at 2,400 feet over wide bands of snow and black bog on the summit plateau. Skylarks were singing at this height and in the incredibly still air golden plover yodelled and called plaintively 'tlui' against a background of nothing! There was no sound of wind sighing in the heather, no noise of aircraft or traffic, train or industrial machinery, no sound of human voices. The natural world seemed reborn. All around us were the sharp crows and barks of red grouse seemingly telling us to 'go-bak' from their territories. In this region lies the National Nature Reserve of Moor House – all ten thousand acres of wet 'desert' – with a thick blanket of boggy peat in some places up to twenty feet deep, which overlies the glacial drift and parent rock. Here the annual rainfall is anything from sixty to a hundred inches, snow cover lies for eighty days in the year and there are below-freezing temperatures in every month, The rain that has helped to create the peat bogs also causes their erosion. If you look very carefully in the channels of eroded peat you can sometimes find the roots of birch trees that grew here eight thousand years ago. Studies are carried out here of the natural fauna and flora and also into sheep production, water storage and tree cultivation at this great height.

From Great Dun Fell a journey across a swampy depression and up a 400-foot slope brings you to the summit of Cross Fell, the highest point of the Pennines at 2,930 feet. I first climbed it in 1943 from Kirkland and I have not forgotten the view from the top on that summer's day. There was a panorama of northern England west across to the Lake District, north to the Solway Firth and Cheviot, south to Mallerstang, Ingleborough and Pen-y-Ghent and east over the hills to industrial Durham and towards the North Yorkshire Moors National Park, with its broad-leaved woods and State Forests, Early Warning Station and nature trails, Rievaulx and other medieval gems. As I sat on Cross Fell among the fluttering, staggering craneflies and listened to the sounds of the moorland birds, whose lives I had shared and enjoyed from the Peak onwards, and the soft sigh of the hill breeze in the wind-trimmed moss, I knew that this land of fell and rocky spur, crag and scree, grey snakelike walls and brown farmsteads, green pastures and rare mountain flowers, stirring wild voices and the drone of insects was one of the most precious assets not only for me but for all of us who love solitude and high wild places.

12. North-East England and the Borders

The old Saxon Kingdom of Northumbria at its zenith stretched from the Humber to the Forth and it is across this broad sweep of English and Scottish countryside that I propose to roam in this chapter. It is a vast region scenically rich and diverse and it is only by pin-pointing those features that have had a special appeal for me that I can hope to reflect its great variety. I cannot describe in detail all those places which deserve mention, such as Spurn Head with its Observatory and scarce migrant birds, or the vanishing land of Holderness, the holiday resorts of Bridlington Sands and Filey Bay, while the white cliffs of Flamborough were featured in Chapter 4. There is the attractive valley of the Derwent, which rises on the Yorkshire Moors and flows for most of its journey along the floor of an old glacial lake. One of my favourite stretches is the Forge Valley, now part of the North York Moors National Park. Steep limestone slopes carry mixed woods and dense undergrowth and I have found wood vetch, columbine, lily-of-the-valley, bloody cranesbill and may lily; in these woods field mice are common and redstarts and pied fly-catchers are active members of the dawn choruses which embrace many thrushes, blackbirds, robins and wrens. There are also out-crops of limestone packed with fossil cockles, oysters and other bivalves; I found a large fossil gastropod six inches long and there are also fossil starfish. The Derwent flows through beds of ramsons, red campion and wild arum and then between overhanging banks and fields, and here water voles, untroubled by human disturbance, swim freely into the river to feed and then play along the water's edge. There is Scarborough with its narrow steep passages called

The Bolts, its massive donjon built by Henry II, its herring gulls nesting on the house roof and kittiwakes on the cliffs above the coast road.

Yet for me the most appealing section of the coast of north-east Yorkshire runs north from Ravenscar – site of a former Roman watch-tower – to Captain Cook's beautiful village of Staithes. It is very different from the coasts to the north and south and is largely made up of rocks of Jurassic age – shales and sandstones. The cliff scenery is dramatic, with a fine bluff at Ravenscar and open moorland or fields of arable land reaching to the cliff edge. One June day I walked along the headland which casts a protective southern arm around Robin Hood's Bay. Through drifts of cold mist, blowing in fine grey strands from the north-east, I could see the Bay below me bathed in warm sunshine. A redpoll twittered over my head and a skylark rose up in song through the mist and into oblivion. From here I made my way to the Nab so that I could walk at low tide on the sun-warmed shore. I picked up several pieces of jet – fossilized driftwood black and feather-light, two fossil *gryphaea* or 'devil's toenails' and three ammonites washed out of the rocks by the sea. A young jackdaw called for food from a ledge above me and, as I looked up, suddenly there was a trickle of pebbles down the cliff followed by a sickening thud and a huge sandstone boulder fell on the spot where I had just been walking, sending up a small cloud of acrid yellow chips and dust – a reminder of the instability of this part of the Yorkshire coastline. I then carried on over the moor near Fylingdales, with its vast silver spheres of the Dewline system and sad little corpses of lambs killed by unthinking heartless drivers, and down the steep one-in-four road that tumbles into the little village of Robin Hood's Bay, with its terraced alleyways and cobbled passages. Swifts were screaming in their spring ecstasy around the rooftops and diving low over the holidaymakers making their way down to the slipway and the beach. Just around the corner, past the sea-urchin stall and the patient ponies with their youthful riders, I found some sand martins flying back and forth in front of a high cliff of boulder clay left by the ice sheets on top of the Lias and now riddled with nesting tunnels.

As the tide retreats it leaves flat rocky pavements exposed and these are known locally as 'scars'; there are pools of seawater left on these raised reefs and here I have amused myself catching starfish, blennies, butterfish and edible and shore crabs. Herring gulls sit on the scars in long thin lines of white and then suddenly the sea mist blows in, the temperature drops and the beach is soon deserted. On such a day I move north on to the moorland plateau and then to

Whitby, with its ruined abbey where Caedmon came to sing, its port with red-roofed houses reaching upwards from the quay and the harbour, where shags and sometimes even guillemots are often busily fishing. From Whitby I have travelled on several times to the little village of Sandsend. Here before the Ice Age one large river entered the North Sea but the boulder clay has now created two streams, separated by Mulgrave Woods which in summer are flushed yellow-green with that umbellifer of the seaside – alexanders. One summer a sedge warbler sang – the first according to local opinion for a decade – and I went on to walk under the dripping cliffs beneath the old railway line and picked up more ammonites – this time brilliant with fool's gold – as well as pieces of blood-red jasper and warm glowing carnelian. Pigeons were nesting above the cliffs and caves and all the way along I could hear the eerie cackle of fulmars. The old railway line was a natural wild garden a-glow in the evening sun with golden clumps of gorse and birdsfoot trefoil; it was a paradise for flowers of the pea family and I also saw red and white clover, hairy tare and common, wood and tufted vetches. Linnets and yellowhammers loved the warm banks, a pheasant was calling explosively, a party of starlings came to roost in some willows on the embankment and groups of herring gulls, sitting or nesting on the bare deserts of the ash-grey clifftops, complained of my disturbance of their peace – the only querulous note on a perfect June evening in north-east Yorkshire.

Just around the corner on the coast is industrial Teesside with its petro-chemical, steel and engineering works and its Seal Sands, which have been said to possess all the charm of the biggest rubbish dump in the word. Yet Teesmouth is one of the top twenty estuaries for wading birds in the British Isles. In August 1962 the pollution of the Tees resulted in the death of thousands of sprats and this event was followed by some truly remarkable ornithological scenes – flocks of thirty thousand fulmars, six thousand kittiwakes, more than fifty thousand other gulls, two thousand terns and several hundred Arctic and great skuas. A few miles farther still is the tidal River Tyne and kittiwakes – those demure nesting gulls of sea cliffs – nest on the fronts of warehouses as far upstream as Newcastle and Gateshead and are well worth seeing.

One of my most regular journeys over the last twenty years has been by road from Scotch Corner, over the Roman Wall to Edinburgh and beyond. This is the course of the A68 – the line of the Roman Dere Street – which offers to the traveller a most rewarding transect across northern England. The route starts as the B6275 through Piercebridge – an attractive village with a green on the site of a large Roman camp – and once the most northerly limit for the nuthatch, whose present breeding range has been pushed north to beyond Alnwick. The A68 runs over the magnesian limestone, the Coal Measures, the Millstone Grit and the Carboniferous limestone; open rolling wooded country gives way to bleak moorland and slag heaps, bare uplands and again to fields and woods. The journey takes me through Royal Oak and West Auckland – part original eighteenth-century houses, part restored – set around one of the large greens typical of Durham and said to be survivors of the compounds into which cattle were driven for protection during Border raids. Pied wagtails and jackdaws feed on these greens.

The northward trail carries on through Fir Tree, where I sometimes pause for a substantial lunch, and up to Tow Law with wide views across rolling country. Ahead is the great iron works of Consett like a giant metal fortress creating in daylight its own miniature white cumulus clouds and at night sending out long red flames that lick upwards into the darkness. To the left of the road is the Derwent Reservoir, which has mallard and sometimes whooper swans and wigeon. Then the A68 runs on to the Carboniferous limestone and descends steeply to the valley of the Tyne with Hexham to the west and Corbridge dead ahead. My first visit to Hexham was in the month of July, and I stayed close to the walls of St Wilfrid's great abbey. The sky was a dull lead and there was a keen freshness in the air; above distant Cross Fell the Helm Wind was blowing. All round the abbey echoed the persistent cries of young birds calling insistently to their overworked and shabby parents – young starlings, house sparrows and blackbirds. Rooks cawed in the trees and newly fledged jackdaws sat on the abbey tower or in the trees in Beaumont Street, their voices less incisive than those of their parents but no less recognizable for all that. Jackdaws also roosted in the trees above the roadway to the detriment of my car roof. As they left in the morning so the swifts began their low-level chases above the houses of Hexham and the site of the ancient cloisters.

The abbey stands squarely in Hexham; its eastern front with six lancet windows rises up above the market square and its gay candy-striped awnings. Inside the abbey are St Wilfrid's Chair – the Frith Stool – carved from one piece of stone, a large memorial stone to a young Roman standard-bearer, the Night Stairs down which the abbey choir processes at evensong and the Saxon apse containing the coffin of St Acca, who succeeded Wilfrid in 709. All these treasures I once described in a BBC television film, and I was particularly impressed with the crypt built of stones brought from the Roman camp of Corstopitum and revealing carved olive leaves cut by a mason in the mists of north-east England who was perhaps pining for the sun and olive groves of his Mediterranean home. Certainly Hexham reveals the significance of stone used to express man's faith in the future and his record of the past.

The nearby Roman Wall marked the northern frontier in Britain of the Roman Empire. The Romans built roads earlier than the Wall, for Julius Agricola, the Roman Governor of Britain, had by AD 84 or 85 conquered northern England and the Borders, and built a system of strategic roads and forts. However, the Romans were forced largely to evacuate Scotland in the early part of the second century and the Emperor Hadrian began the construction of a wall across the narrow part of England from Newcastle to the Solway Firth – a distance of eighty Roman miles or seventy-three modern ones. The original height of the wall was about fifteen feet, with a parapet of some six feet on top of that; its width varied from eight to ten Roman feet. On the north side was a great ditch, or fosse, and at every mile interval there was a fort known as a 'milecastle', and at certain strategic sites on or near the wall were forts with garrisons. South of the Wall was a ditch called the 'vallum', with an earth rampart on each side.

I have walked many miles of the Wall itself – and you will never know the magnificence of it without doing so. There is the crossing of the vallum near Benwell at the Newcastle end, and I have travelled along the metalled section of the old military road built after the 1745 rebellion, and visited the old section at Brunton, where in the nearby wood willow warblers, redstarts, song thrushes and blackbirds were singing near the Turret. The Wall crossed the North Tyne – by means of a stone and timber bridge, and it is still possible to see the great abutment of the bridge near Chesters, also set among trees, where great tits and robins sing their territorial songs. On the west bank was the cavalry fort of Cilurnum – Chesters – set in wooded parkland, and here the ruins reveal the remains of the Governor's house, while the bath house, set between the fort and the river, overlooks the watery haunts of common sandpipers and grey wagtails. The museum at Chesters has sculptured reliefs of stags in a forest or being hunted by man. The walls of Chesters are home for several interesting plants such as yellow Corydalis, shining cranesbill and fairy foxglove; the last has been called 'the Roman Wall plant' but it was apparently placed there by a vicar of Stamfordham who wished it to be thought a relic of the Roman occupation.

Beyond Chesters there are fine views from the Wall of the North Tyne valley and the Cheviot Hills and here the vallum and wall have

184

been cut through a band of dolerite. Then the Wall begins to follow the dolerite ridge of the Whin Sill, which we last met at High Force and High Cup Nick. This stretch has been called 'Wall country' since it is formed from moorland, where curlew and lapwing, skylark and meadow pipit breed, and several lakes, known in the region as 'loughs' – pronounced 'luffs'. This was once a wild and dangerous countryside – the scene of cattle-raiding and the haunts of Border robbers. At Sewingshields it is believed that Arthur and his knights lie asleep, while Housesteads is the best preserved of the forts, with ramparts, granaries, barracks, latrines and stone roads with deep ruts from the passage of countless chariot wheels. Set in rolling countryside Housesteads is impressively sited and looks across to Vindolanda, where writing tablets have been found, and in a lavatory gold coins in a purse. The three loughs of Greenlee, Broomlee and Grindon lie on a flyway to the Solway from the east coast of England and are worth inspection in the winter; I have seen mallard, wigeon, teal, pochard, tufted duck, goldeneye, goosander, black-backed gulls and mute and whooper swans.

Farther to the west at Winshields just short of the highest point on the Roman Wall the view unfolds over the outcrop of the Whin Sill, Crag Lough and mile after mile of open country with grazing stock. On a clear day one can see the hills of south-west Scotland and the Lake District. Between Thirlwall Castle and Housesteads the Pennine Way follows the line of the Wall past Crag Lough and Cuddy's Crags and between Greenlee and Broomlee Loughs before passing through Forestry Commission plantations full of coal tits and goldcrests, and so on to Bellingham. On the Wall one of my favourite places is the crossing of the River Irthing, which was controlled by the Romans from the fort at Birdoswald; a nineteenth-century Earl of Carlisle thought that the view across the valley was the finest in Cumberland and he compared it to the view at Troy. Skylarks forge a chain of song above the fort and blackcaps, willow warblers and wrens sing in the woods near the old bridge abutment, but the river has changed its course somewhat since Roman times. At the River Irthing the scene becomes less wild as the limestone belt comes to an end and the red sandstone of the Irthing and Eden Valleys begins. For years the Wall, snaking along the Whin Sill and meandering across the landscape, has been plundered for building stone and only nettles or small woods and modern roads mark its ancient course. In the west the remains of the Wall peter out towards Carlisle and the Solway. To have walked parts of its length is to have shared something of the experience of the Roman legionary, far from home, staring northwards into a misty and hostile land.

185

The Roman Dere Street runs past Colt Crags Reservoir and here there is a choice of travelling on to Rochester or west towards Bellingham and Kielder Forest. The Border Forest of the Forestry Commission covers one hundred and eighty square miles and is perhaps the largest man-made forest in Europe; it is composed largely of Sitka and Norway spruce. When I carried out censuses of birds in Kielder Forest some years ago I found that the Sitka spruce plantations were dominated by coal tits and goldcrests in about equal abundance but on the whole the coal tits showed a greater preference for Norway spruce. Blackbirds, greenfinches and carrion crows could be seen in both kinds of plantation and there were also pheasants, herons, robins, great, blue and long-tailed tits, treecreepers, great spotted woodpeckers, mistle and song thrushes, linnets and yellowhammers, while on the open moors and ecotones were black and red grouse, ravens, meadow pipits, skylarks, whinchats, nightjars, dippers and wagtails. I also came across badger setts and fox earths, roe deer, red squirrels, blue or mountain hares, but I did not see the wild goats that are said to roam on the fells above Kielder. The region is certainly one of the most extraordinary that I have visited in Britain, and, although I am not over-enamoured of conifers carpeting the hillsides, the wildlife is very interesting. It can be a windy place, as Swinburne recorded: 'On Kielder-side the wind blaws wide'. My alternative cross-country route runs through Ridsdale, above Otterburn, where Sir Harry Percy fought the Earl of Douglas in 1388, through great regions of conifer forest where the roe deer graze by the roadside and so up to the Cheviot Hills, passing Catcleugh Reservoir on the left, for me on most occasions aggravatingly devoid of birdlife.

The great bare domes of the Cheviot Hills provide the most spectacular scenery in the county of Northumberland and they occupy perhaps two hundred square miles of that county and another hundred square miles of Roxburghshire. They also comprise the northern part of the Northumberland National Park stretching from near Wooler to the Irthing, and the Park includes part of the Wall and the end of the Pennine Way. The Cheviots are formed from massive hills of volcanic rock of Old Red Sandstone age; in fact they represent a large extinct volcano which scattered ash over the region and then poured out vast quantities of lava. Now the hills are ice-smoothed and rounded and largely covered with grass; in the real Cheviots there is hardly any bracken or heather. It is good walking country although Cheviot itself rises to a height of 2,674 feet and is covered at its summit range by a bog. From the top there are remarkable views over the rolling hills and the summits of Comb

Fell, Bloodybush Edge, Windy Gyle, Hedgehope and Black Hog. The range is thinly populated and its prairie-like green slopes are largely given over to grazing by Cheviot, Black-faced and cross-bred sheep. Radiating from Cheviot are becks or burns like the Usway, Breamish, Harthope and College, some in steep-sided narrow valleys, which add the gentle music of running water to an already quiet and peaceful range of hills. The slopes are the haunts of red grouse, whinchats, wheatears, ring ouzels and sometimes high breeding blackbirds. There are grey wagtails and dippers on the streams which are full of mayfly larvae, stonefly nymphs, various fly and beetle larvae, water bugs and snails. There are well-wooded valleys like Langleeford, where I have rejoiced in the songs of redpolls and redstarts, pied flycatchers and tree pipits, while above the tree-line there are meadow pipits, wheatears, ravens, ring ouzels, golden plover and perhaps even merlin and a nesting pair of dunlins. These higher slopes of the fells are worth ascending not only for these upland birds but also for mountain plants such as cloudberry and dwarf cornel and bog plants including saxifrage, butterwort, bog asphodel and grass of Parnassus; there are also wild goats, foxes and both kinds of hare in these high regions. There is often a peace and a comforting, brooding stillness about these grassy and boggy places – a contrast to the river valleys like that of the Breamish, bright with monkey flower and foxglove and full of the songs of small birds. In winter Cheviot may lie under swaths of snow and I remember driving one December day in a growing blizzard up to Carter Bar and as the snow accumulated the countryside turned to stark black and white. Cheviot looks down on two noteworthy villages – Glanton, from whose Bird Research Station long-term studies have been made since 1933 of the dawn choruses of the birds, and Chillingham, with its herd of wild white cattle, descended from the wild ox, which roam in natural surroundings of wood, fell and crag.

From the summit of Cheviot one can also look across to the coast of Northumberland and the Farne Islands. This is one of my most loved stretches of British coastline; there is the noble sight of castle and priory in Tynemouth, of Coquet Island – the retreat of Henry the Hermit and the breeding haunt of many terns – of Amble with its cobles under construction and its port, of Howick Hall with its fine magnolias and rhododendrons, of Craster, famous for its kippers, and the Whin Sill which once provided London with stone. There are also Dunstanburgh Castle, which has been a royal castle since about 1400, with its Lilburn Tower standing on a cluster of pillared rocks, Seahouses, Bamburgh, the Farne Islands and Lindisfarne – for me all places of interest, history and magic.

My birdwatching base in 1959 and 1960 was an old grey rambling house of infinite charm, nestling behind great sand dunes with its own bay and a splendid view of the low dark islands of the Farnes. This was Dr Eric Ennion's Bird Observatory at Monk's House, halfway between the delightful harbour of Seahouses and Bamburgh, once the capital of Northumbria. This 'house on the shore', occupying a site on which buildings have stood since the thirteenth century, thought Eric Ennion, 'as well as a granary and ferry house for the Farnes, also comprised a graveyard and a chapel'. There are nearly always birds to be seen here and fossil crinoids – St Cuthbert's beads – to be picked up on the shore. The lighthouse beams from the Longstone used to send a patch of yellow light circling slowly round my bedroom ceiling. Along the road from Monk's House to the north in the summer corn buntings had established territories, and in the evening the cock birds sang their key-rattling ditties from the telegraph wires before gliding down, legs a-trail, to the ground below. West of the road in some marshy ground reed buntings had their nests but some of the males had extended their territories into the sand dunes on the seaward side of the road. One male had a favourite song post on a tuft of marram grass and here, resplendent with coal-black head and snowy collar, he used to sing and watch me, look inland towards the rushy meadow or out towards the Farne Islands. In winter this pool was visited by other birds. One November there were scores of gulls and redshank and smaller numbers of mallard, wigeon, teal, shoveler and one superb drake goldeneye. The low grunts of the feeding shoveler mingled with the sharper whistles of the wigeon and mallard. In January 1965 I saw a first-year glaucous gull from the Arctic on this pool – sandy in colour with whitish wing tips and a dark tipped bill with which he regularly stabbed off any gulls that approached him too closely.

Monk's House was also an interesting place to see migrants in the autumn – redwings and fieldfares, barred and garden warblers, siskins, goldcrests, redstarts, bramblings, crossbills and even such rarities as the lesser grey shrike and Pallas's warbler. The beach too is frequented by waders like sanderling, turnstone, little and Temminck's stints and curlew sandpipers. The rockpools not far away are full of interest, with limpets and mussels, acorn barnacles, periwinkles, shrimps and prawns and red beadlet and dahlia anemones. There were butterfish and sea scorpions, common starfish, beautiful sunstars, and delicate brittle stars all of which I found in a tiny rocky gully with sand at its head and turf-covered dunes behind. Giant sea slaters crawled over the dry rocks; clad in

188

grey armour they were fearsome-looking animals. In the dunes cinnabar moths frequented the ragwort and song thrushes came to look for banded snails to hammer out on their stone anvils.

Just up the road from Monk's House is the ancient township of Bamburgh, with its mighty castle lodged on an outcrop of the Whin Sill and founded by King Ida in AD 547. For me it is perhaps the most impressive and dominating castle anywhere in England. One May morning I was out at four o'clock. The morning air was fresh and cold beneath the great volcanic cliff on which the castle stands. In the distance I could hear the North Sea breakers falling intermittently on the sands, where some of the most memorable scenes in the film *Becket* were shot. Just at half past four the first bird began to sing; it was a song thrush proclaiming his occupation of a territory to the northern dawn and a minute or so later he was joined by a blackbird singing from a hedge near the cricket field under the dark, brooding castle walls. At five o'clock some early rooks from the rookery on the green in the middle of Bamburgh not far from St Aidan's Church, where Grace Darling lies at rest, added their raucous caws to the now swelling dawn chorus, and wrens and robins also joined in. Under the castle crags the brambles and ivy clinging to the stone began to murmur with the stirring starlings and house sparrows adding, as they woke up, their whistles and chirps to the chorus of sound around the castle.

My chief interest in Bamburgh Castle, however, was in the fulmars – those ocean-going members of the petrel family – which gather and nest on some of the volcanic ledges. Fulmars first prospected this aery site in 1931 and an egg was taken in 1934. Since then fulmars have bred regularly on the castle crags. At first I did not see many birds present since most were still out at sea but at twenty past five they began to appear, flying and gliding on pearly-grey wings with consummate ease and giving their quiet, little-known grunts. In the east the sun was now rising as an almost round moon dropped away to the west. The light from the rising sun illuminated the castle towers and caught the undersides of the outstretched fulmars' wings with a golden glow. Birds now flew towards the ledges, either to hover for several seconds and drop away, or pitch rather inelegantly on the ledges or in the rock crevices. The characteristic cackling of the fulmars now began to echo round the castle. Some of the birds were disputing the ownership of cracks in the castle's bastions and others were embroiled with noisy jackdaws wanting the same site. Above the castle woodpigeons, starlings, swallows and linnets were migrating northwards and, as I turned away from the cliff, a whimbrel, perhaps on its way to Shetland from

189

Africa, passed overhead, giving one of the most stirring of all wader calls – the whistles from which its name Titterel or Seven Whistler are derived. With this melodious sound in my ears this early May morning was complete.

Let us now envisage a group of rocky islets, encompassed by a grey and forbidding sea, whose restless margins swell and beat remorselessly against the rocks! Thin wedges, these stony volcanic islands of the Farnes – from fifteen to twenty-eight in number according to the state of the tide – and lying off the Northumberland coast are the eastern outcrop of the Whin Sill of which we have heard so much. Through the kindness of the Farne Islands Committee of the National Trust which owns the islands I have been privileged to study the birds and seals on this romantic group in the North Sea. I first visited the largest island – the Inner Farne – in July 1955, and having been caught in a sea mist was forced to spend the night with a lost homing pigeon in the medieval St Cuthbert's Chapel. According to the Venerable Bede 'No one, before God's servant Cuthbert, had ever dared to inhabit the island alone', and it was here in AD 687 that the former shepherd boy from the Lammermuir Hills but now Bishop of Lindisfarne died – one of the greatest figures in the early Christian Church.

On several July mornings in the 1950s and 1960s the coble *Glad Tidings* from Seahouses, in the capable hands of Billy Shiel, carried me to the Inner Farne – sixteen acres of islet, formed of rocks and cliffs where wind and storm have gashed the edges of the land into contorted pinnacles and faces. At one spot where the cliffs are nearly eighty feet high there is a deep crack called The Churn, and here storm-driven waters press wildly up a shaft and may reach a height of ninety feet in a northerly gale. The landing place on this hallowed sanctuary is in St Cuthbert's Cove beneath the medieval chapel and the fifteenth-century tower of Prior Castell. Here myriads of Arctic terns on snow-white wings and with harsh screams objected to my invasion of their island refuge. As I made my way across the seaweed-covered rocks and up on to the short turf of the Inner Farne, bright with thrift, sea campion and *Amsinckia,* I could see scores of tern chicks of every size, from little fluffy brown balls to free-flying juveniles. Some of the Arctic terns began their dive-bombing attacks on my head, drawing blood with their beaks and uttering penetrating screams and sharp rattling calls like diminutive machine-guns.

Below the lighthouse which crowns the highest point on the island there were steep cliffs and gullies where kittiwakes had built their crowded, noisy cities. There were razorbills and guillemots nesting in cracks and on ledges in the Whin Sill, shags were nesting nearer the sea and belched at you in alarm if you peered over the clifftops, and puffins assembled on a flattened gathering ground among the sea campion. Eiders – St Cuthbert's own ducks to which he gave complete protection – were sitting on downy nests along the lighthouse wall and near the chapel. It is the terns, however, which dominate the island in the summer months – Arctic, common and large crested Sandwich terns – and the rarer roseate which may even pursue and harass the large Sandwich terns. The roseate is more slender than the Arctic tern and has conspicuously long tail streamers and a unique combination of red legs and black bill. Its rose-pink breast, which led William Darling, Grace's father of the famous *Forfarshire* rescue, to describe it in his notes as 'the pink-breasted tern', has often disappeared by mid-July. Its voice is a help to identification – a series of harsh 'aaaaks' or a short clear 'chew-it' but I had to make four visits before I could tape-record the sound clear of all the other tern noises. I also visited Staple Island in summer to see and film the guillemots massed together on the Pinnacles – flat-topped isolated blocks of the Whin, rising to almost sixty feet in height.

On the south and west of the Farne Islands are the steep cliffs where the seabirds have their homes, while on the north and east are sloping beaches, and here on these dark slopes in November is unfolded the mysterious breeding cycle of one of Britain's most appealing and fascinating mammals – the Atlantic or grey seal. The great bulls haul up to six hundred pounds of bone, muscle and blubber onto the Brownsman, Staple Island and the North and South Wamses. I have been fortunate to stay in the cottage on the Brownsman to which in 1815 came William Darling for a stay of eleven years before he moved to the Longstone. The Brownsman is smaller than the Inner Farne – about ten acres in size with extensive areas of rock, a covering of peaty soil and flowering sea campion. I lived for a week with the seals, watching the bulls haul up around my doorstep, take up their territories and be joined by the cows, which gave birth to creamy yellow pups and mated again before returning to the sea – the whole of the breeding performances being telescoped into three weeks. I could hear the pups crying plaintively to their mothers and the cows replying with long mournful wavering 'songs' – but not quite those that the Sirens sang! If I went too close to a pup on its back it would roll over and hiss angrily as lubricating tears ran down its face. I watched a pup being born and another being given swimming lessons at three days of age.

During my November stay on the Brownsman I was also able to witness the arrival of migrant birds from Europe, including

191

fieldfares, redwings and blackbirds, starlings, chaffinches, bramblings, skylarks, woodcock, snipe and two parties of six and eleven herons. There were a merlin, a sparrowhawk, and a peregrine hunting the tired migrants. From my vantage point on the island that November I also saw a killer whale hunting seals, and on the sea such interesting birds as great northern diver, red-necked grebe, velvet scoter and up to fifteen little auks. My visits to the Farne Islands have always been full of interest but the seal population has grown and culling has become necessary to preserve the quality and health of the stock and to stop the overcrowding and degradation of the environment. Seals are attractive mammals to us and the arguments are still carried on about culling, as I found out when I came to make a feature on seals for the BBC's 'Woman's Hour' programme. To live amongst them was a great thrill and when I left the Brownsman in the *Glad Tidings* just before a south-west gale blew up I was surrounded by adult seals, which looked up from the sea like mournful dogs and serenaded me with their eerie songs. The Farne Islands are sometimes misty and remote and sometimes there is the wail of the Longstone for company but always they have been enchanted islands for me.

The wild calls of wading birds and wildfowl are very much sounds that I associate with an area of tidal flats to the north of the Farnes; these are guarded by Holy Island (Lindisfarne) and the dunes and sands of the Old Law, with its twin pinnacles of sea-marks. One cold November day I stood on this great bight of mud flats and I could see to the west the whale-back of Cheviot streaked with recent snow and to the east the sheer bluff of Beblowe Crag. A small skein of pink-footed geese made their way high up over the Slakes and I was reminded of those words of Shelley – 'boundless and bare, the lone and level sands stretch far away'. I moved south-east to Chesterhill Slakes and Budle Point on the southern side of Budle Bay. Here in bright sunlight I gazed on literally acres of wigeon with countless gulls and many mute swans, mallard and a few shelduck. Some of the shelduck were stalking about the flats with a quiet aldermanic dignity, others were sieving for food in the mud and some were flying over in a flurry of white, black and red, laughing deeply as they went. Two young men in green anoraks walked along the northern bank and as they made their way around the Bay the mute swans made off in a stately progress away from the disturbance, and then the wigeon began to fly up, the air becoming alive with the musical throb of the swans' wings and the lovely slurred whistles of the wigeon.

North of Budle Bay are the Ross Back Sands, the open crescent of Fenham Flats and Holy Island Sands, where traditional shooting is allowed by permit – and this was once great punt-gun country – while the rest remains a sanctuary within the Lindisfarne National Nature Reserve and the main refuge for migratory and wintering wildfowl in north-east England. Wigeon are common here but I have watched whooper swans and flocks of greylag and pink-footed geese, while the little black brent geese for whom Lindisfarne is the only regular wintering area in Great Britain are mainly of the light-bellied race. Eel grass or *Zostera* is an important attraction for many of the wildfowl. In winter this is a fine region to see wigeon and eider and also scoter, scaup, mergansers, goldeneye and long-tailed ducks that 'yodel' in a most musical way producing, as Richard Perry who studied them off the Gull Bank on Holy Island wrote, a 'medley of notes, variously broken or modulated by wind and sea, that produced the effects of bagpipes!' Then there are the wading birds – curlew and godwit, sanderlings, turnstones, ringed plover, dunlin, purple sandpipers, lapwings, oystercatchers, golden plover, grey plover and knots – some of them veritable globe-trotters and bringing to the flats the music of wild places and a reminder of far-off lands.

The causeway to Holy Island from Beal is for me a magical gateway where my sense of history heightens my appreciation of the mystery and character of this somewhat inaccessible island. From the mainland one sees a low island of dunes dominated by a flat panorama of houses and the remains of the Priory, while Lindisfarne Castle perches on its cone of rock at the southern end. On Holy Island I have marvelled at the wild flowers – red clover fields alive in summer with blue and brown butterflies, strands of lady's bedstraw and golden birdsfoot trefoil in the grazings, and on the Whin Sill of Beblowe Crag clumps of mallow, valerian, thyme, stonecrop, wallflower and even henbane. The sandhills are habitats for hound's tongue, which I also found at Bamburgh, centaury, gentian and purple orchid, and here the skylarks and meadow pipits sing all day. Viper's bugloss raises purple spikes along the banks, where wall butterflies and tortoiseshells bask or flit to and fro. All the way from the Bibly Hill to the Snook the larks go up in song. I have walked all around the island on a May day and have never for a moment been without the song of a skylark pouring down on me from above. On the Lough moorhens call amongst the bogbean and the amphibious bistort's spikes of flowers and one day I counted more than five hundred black-headed gulls. On the Coves I watched thirty fulmars nesting on the rock face, and in the thorn hedges of the Straight and Crooked Loanings I have seen late fieldfares, linnets, greenfinches, redstarts, willow warblers and Greenland wheatears. During the

193

seasons of migration there are often spectacular falls of migrants and sometimes very rare visitors indeed.

The ultimate glory of Holy Island is Lindisfarne Priory standing among its meadow saxifrages, one of the most significant monuments in England. The isle of Lindisfarne was given by Oswald to Aidan, the missionary bishop sent from Iona. Apart from a few carved stones little remains of the original monastery and the period when Cuthbert was bishop; the best representation of that time is provided by the seventh-century Lindisfarne Gospels now in the British Museum. Today there are just roofless square towers with round arches or slits, fragments of pier arcades and arches, and a dramatic diagonal rib that springs from one crossing pier to another with chevron mouldings, and is known as 'The Rainbow', all belonging to the period after the Norman Conquest. It is a place of inspiration but my favourite journey is down to St Cuthbert's Isle on Hob Thrush and here, on the grassy basalt rock which is accessible only at low tide, I can sit within the foundations of a little chapel that according to Bede was used by St Cuthbert as a retreat. This is one of the most sacred places in England and from its sanctuary I have watched the eiders – the Saint's own ducks – swimming past. Holy Island that figured in Scott's *Marmion*!

> For with the flow and ebb, its style
> Varies from Continent to Isle;
> Dry-shod, o'er sands, twice everyday,
> The pilgrims to the shrine find way.

Inland the Borders reach towards the Firth of Forth. One route takes me up through Jedburgh – once scene of almost continuous bloody fighting – with its warm sandstone Abbey founded by David I and with a long rather stark nave where the jackdaws gather and preen. Great pink outcrops of the Old Red Sandstone hang over the wooded valley of the Jed Water whose headstreams lie in Wauchope Forest. I once heard a pied wagtail singing on the house where Mary Queen of Scots lodged in 1566 and from which she rode to visit the wounded Earl of Bothwell and returned almost to die in Jedburgh. North-west lies Melrose Abbey among its gravestones and ox-eye daisies – a ruin of great beauty and charm and best viewed 'by moonlight', according to Sir Walter Scott, who lies in nearby Dryburgh Abbey. It rests on the lower slopes of the conical Eildon Hills, old groups of volcanoes, with heathery peaks in a largely grassy upland region. From here I sometimes drive on to Peebles to walk along the river there among the house martins and swifts, the mallard and moorhens; one severe winter I saw starving robins, blackbirds and blue tits looking for scraps in the snowy gutters of the main street. Then Edinburgh is my next stop. But that is only one way to cross the Borders. Other routes have taken me through Hawick and Ettrick Forest, the Lowther Hills, which rise to over two thousand feet, the Moorfoot Hills above the Midlothian coal basin, and the Pentlands of folded Silurian and Old Red Sandstone volcanic rocks.

Another route is to follow the River Tweed to Berwick. I have been a guest at the Hirsel near Coldstream, and here I have recorded some of the sounds of the duck on the loch. One October I hid by the lochside and was rewarded with close views of an old dog fox that loped past my hiding place; on the water there were several hundred mallard and smaller numbers of wigeon, teal and goldeneye. On other occasions pintail, shoveler, pochard, tufted duck and goosander all found the loch a refuge and feeding place. In the summer the woods at the Hirsel are rich in birds. I have seen and heard up to eight pairs of pied flycatchers, three pairs of turtle doves, stock doves, woodpigeons, goldcrests, chaffinches, wood warblers, blackcaps, garden warblers, redstarts, marsh and coal tits, great and blue tits and cuckoos in the Hirsel woods; the richness of woodland birds was one reason why the Hirsel was chosen for the annual England-Scotland radio birdsong contests in the early 1950s in which I once shared a part at the Hirsel. Long-eared owls and tawny owls, great spotted woodpeckers, curlews, grey wagtails, sedge warblers and a single hawfinch are also on the list of birds. Above Coldstream is Fala Moor – a heather moor where large numbers of pink-footed geese used to come to roost among the many red grouse. I joined Peter Scott at Greenlaw to see the rocket-netting of some of the geese one October – birds that had come from Iceland to spend the winter in the Borders. It was a magnificent sight to watch the geese send over a scout to reconnoitre the stubble field and finally drop down; it was followed by small parties and then large numbers of geese moving in long skeins and landing with backward beating wings in a chorus of resonant notes after dropping out of a rose-pink sky into a land of chequered purple and gold.

Again it is necessary to push north past the goldeneye that feed in the Tweed below the Royal Border Bridge, past Tynemouth – refuge for geese and ducks, past North Berwick Law – G. K. Chesterton's 'Hill Absolute' and a volcanic neck just like the nearby Bass Rock with its ten thousand pairs of gannets, past the old Douglas fortress of Tantallon, through the charming village of Dirleton with its Anglo-Norman castle, through Prestonpans with its Jacobite association and finally to Edinburgh.

Beyond the Lammermuir Hills and the Firth of Forth is Loch

Leven – rich in history and one of the largest and most important of all Britain's inland wildfowl refuges. It really lies beyond the Borders but mention must be made of it and there is probably considerable interchange between it and the sanctuaries in Northumbria. Those who drive along the M90 motorway from Edinburgh towards Perth may have noticed near Kinross this loch lying on the righthand side of the road. It has an area of more than three thousand acres, a shore line of eleven miles and some very fine trout. In winter the frost often causes the loch to freeze and the resulting sheet of ice sends out intermittent and resonant sounds – BOING. Castle Island rises from the loch and and here in this dour fortress Mary Queen of Scots debated with John Knox in 1563; four years later she was to be a prisoner within its walls. Closer to Vane Farm and the Loch Leven Nature Centre is St Serf's Island – low, flat and grassy, where large numbers of duck breed in summer including mallard, shoveler, gadwall and pintail, while wild geese roost there in the winter. As long ago as 1629 a visitor to the loch reported that 'there be great store of all kinds of wildfowl, of wild geese and swans many'. The RSPB's Centre at Vane Farm gives a good view of the loch and the fields where the geese often feed.

One December day of heavy snow and frost I made my way with recording gear through the snow down towards the loch. There were mallard, wigeon and goldeneye on a small patch of unfrozen water and a solitary redshank complained from the edge of the loch. I sheltered under a drystone wall which gave protection to me and rabbits, wrens and hedgesparrows. A lone goose inspected the carrot field behind me and returned to St Serf's Island. A few minutes later a party of pink-footed geese with one or two greylags, reassured by their scout, came towards me and called as they passed overhead before dropping down with heavily beating wings into the carrots. Their wild stirring calls in a bleak and snowy landscape were for me the very breath of winter in the Scottish countryside. This is a land of lochs and hills – inspiring scenery both in summer and winter, that bridges the lower hill ranges and moors of England with the Highlands and old Caledonian Forest which begin to unfold north of Loch Lomond in the west and Perth in the east. Here beyond the Borders we stand on the brink of a new and exciting excursion – deep into the Scottish Highlands.

13. Bottle-Green Pines

For someone who lives in the twentieth century like me it is rather difficult to imagine what northern Britain may have looked like some twelve thousand years ago. Yet we can deduce that blue-white sheets of ice covered a great deal of the region and in Scotland the ice was at least 3,500 feet thick in the hollow basin in which Loch Maree now lies. against wolves. Cheviot and black-faced sheep were brought in to surrounding glaciers. To the south of the ice lay a strip of tundra where the rocky, gravelly ground was frozen for most of the year; here there were no real trees and only dwarf shrubs, mosses and lichens. This was a land of the ptarmigan, snowy owl, long-tailed duck, knot and snow bunting, while perhaps elk, musk oxen and reindeer grazed over the infertile ground. From about 18000 to 7500 BC there was an improvement in the climate, and the glaciers began to melt with boulders and gravel tumbling down in the roaring streams of melt water. It remained cold to the south of the ice and typical plants were still Arctic birch, willow and mountain avens. The summers then began to warm up and birch trees started to colonize the valleys and flows, following up the edge of the retreating tundra. Not far behind came Scots pine and some hazel, oak, aspen and alder. Slowly the pine began to establish itself and in these early mixed woods deer and wild cattle and large predatory animals roamed and beavers built their dams across the slowing streams.

For quite a long period a large part of the British Isles was clothed with Scots pine but then the forests were felled for timber or burned against wolves. Cheviot and black-faced sheep were brought in to graze the land, and forest regeneration was halted. Persecution brought about the loss of bears, wolves, lynxes and polecats, kites, ospreys and capercaillies – the last were later re-introduced from

Sweden – while wild cats and pine martens only just held on. Some relics of these ancient Boreal pine forests still survive today almost entirely due to their remoteness and the difficulty of extracting their timber. The most significant of these woods, despite being often fragmented and scattered, are to be found in Rothiemurchus, Glenmore and Abernethy Forests in Strath Spey, Inverness-shire, at Ballochbuie and the Forest of Mar in Deeside, on the southern shore of Loch Rannoch in Perthshire, in Glen Affric, Glen Torridon and Glen Shieldaig, at Coille na Glas Leitire by Loch Maree and at a few other isolated spots such as Loch Hourn and Glen Carron. Because of past heavy felling some of these woods are rather open and form a heathland dominated by heather and blaeberry. The old Caledonian Forest is thought to have stretched from Glen Lyon and Rannoch to Strath Spey and Strath Glass and from Glencoe east to the Braes of Mar. The Scots pine was in fact chosen as the badge of the Macgregors, Farquharsons, Shaws and Grants, perhaps to show their association with the great Forest of Rothiemurchus – 'the plain of the great pines'. Birch trees often grow alongside pines since they have similar demands as to soil and climate but the birch can grow nearer to the North Pole. In Scotland the Highland pinewoods probably filled the space between the oakwoods in the bottoms and on the lower slopes and the birches at the highest points. However, some pines reach quite high up the slopes in the Cairngorms and indeed I have listened to a goldcrest singing in low pines at over eighteen hundred feet above the River Feshie.

For me the month of June is the richest and most rewarding of all the exciting months in the Scottish Highland year. Nowhere is this more true than in the Forest of Rothiemurchus in Strath Spey, the great settled valley which runs from the Moray Firth to the Great Glen. Large farms occupy the better soils and valley slopes, while crofts are dotted about the higher reaches, but both are dependent on cattle and sheep. Rothiemurchus lies in a basin with the River Spey and the grey mossy Monadhliath Mountains of metamorphic schists on one side and the high Arctic plateau of the granite Cairngorms on the other. The latter mountains are often streaked with snow and their peaks shrouded in a white mist which hangs forbiddingly above the glens and the open stretches of pine forest. I have lived for weeks at a time by the shore of Loch Morlich which, before the opening up of the area to tourism and winter sports, was a remote sheet of water set in superb mountain scenery. Around the loch grew bottle-green pines in open or close canopy. The young trees were pyramidal or cone-shaped but as each tree matured so the lower branches died and the crown became flat or dome-shaped. As the outer bark fell off, it revealed a reddish shade beneath and the combination of shape, orange-red bole and green canopy made the Scots pine a truly magnificent tree.

As I gazed out from my summer camp I could see common sandpipers bobbing and tittering on wet boulders by the loch or flying in display between the pines, while snipe chippered insistently in the sphagnum bogs among the butterwort, bog bean, water lobelia and bog asphodel. Oystercatchers, gadwall, teal and goosander came to feed close by, chaffinches and black-headed gulls commandeered the camp and only a few yards away were a pair of greenshanks with young. These rare and handsome wading birds – dramatic, fast and melodiously voiced – are one of the supreme attractions of the Highlands. The female used to survey the loch from the topmost shoot of a pine and when she became anxious or alarmed she would call with contralto 'Tchootchoos' interspersed with higher 'Tchip-tchip-tchips'. Then she mounted high in the air to sweep down to within two or three feet of the ground and fly towards the intruder at tremendous speed.

Several times one year I heard the male greenshank's song over the loch and my recording of that lovely sound takes me back at once to that particular June morning with its warmth, peace and solitude. For miles around the loch unfolded the remains of the Caledonian Forest where willow warblers sang all day, whinchats and meadow pipits filled the clearings with their songs and in the trees plump Scottish crossbills 'jipped' and twittered, goldcrests and coal tits wheezed and crested tits with their sharp erectile crests and black gorgets – birds confined absolutely to the pine forests – purred and trilled. The crested tits are insect feeders but during the stalking season they come down to feed on the fat – the gralloch – where the hunters and stalkers have made their kill. The crested tits are one of the great attractions of Spey Side but recently they have extended their range to the Moray Firth, and a small population may have survived in north-west Inverness-shire. Beneath the trees there were capercaillie and black grouse. On a rock buttress that towered above a wooded glen a pair of peregrines had their eyrie and I watched this for three days, witnessing the first flight of the eyasses to their new freedom.

In mid-summer a pastel light often persists from sunset through to dawn and the glens and forest are full of bird song though not as richly as an English oakwood. The main contributors are the chaffinches, willow warblers, goldcrests, coal tits, wrens, redstarts and robins with some tree pipits and cuckoos. It is sheer delight to walk under the resin-scented bottle-green pines, the earth fresh with blaeberry, cowberry, juniper, chickweed wintergreen – a real Arctic relic – lesser twayblade orchid, heath spotted orchid, lousewort, cow-wheat and the very rare twin-flower named scientifically after the great Linnaeus himself. The braes and slopes are carpeted with green-tipped heather, alpine lady's mantle and saxifrage.

As one lifts one's eyes to the mountains there are still perhaps snowbeds in the high corries of Braeriach and Carn Ban Mor and even on Cairn Gorm, criss-crossed as it is now by an eroding maze of human trackways. Clouds of biting midges and fierce clegs assail you as you walk or sit by the loch to watch the ringed plovers and the goosanders. One of the most stirring of all the natural sounds is the bubbling call of the curlew cascading out of the sky and for perfect peace there is a tiny little-known lochan where I can listen to a gentle chain of goldcrests' songs and the soothing plop of trout rising in an ever-widening ripple of activity. One afternoon I watched an osprey – that fine brown and white fish-eating bird of prey which was exterminated in Britain for half a century and returned to nest in Speyside in the mid-1950s. It circled with slow flaps and glides over the loch looking for fish and hovering with sharply angled wings. It was a common sight to see buzzards soaring over the forest; their nests were often in tall pines. Sparrowhawks dashed through the pine forests but I did not see an eagle's eyrie in a pine although this is not very uncommon. Red squirrels scampered after each other and so deeply engrossed were they in their chases that they passed within two feet of my stationary figure and once ran over my boot. Red deer moved quietly through the forest, roe deer barked in alarm at my approach and there were always the wood ants to watch – big columns bringing in plant material and dead flies and caterpillars for distances of up to fifty yards from their domed cities.

At nine o'clock one July evening I set up my recording gear on the edge of the Forest of Rothiemurchus to wait for the roe deer to move out of their lying-up places of the day into the fields for their night's feeding. What little breeze there was was blowing from their feeding area to me so that the sensitive-nosed deer would not pick up my scent. Around my hiding place stood centenarian Scots pines under which straggly clumps of juniper formed an open shrub layer and creeping ladies' tresses orchids pushed up their pale spikes of flowers. Gradually the desultory bird sounds of the evening – tit calls, a burst of siskin song, some willow warbler music – died away, but a single song thrush sang his loud clear phrases until almost ten o'clock. Just after eleven two woodcock began to 'rode' between the trees giving their frog-like display calls, and a small party of oystercatchers came over on their way to the shingle of the River

Druie or the blue lupin-bedecked banks of the Spey. Even after the sun had set the ancient pines still cast appreciable shadows on the ground and I could easily read my watch throughout the night. Then in the stillness I heard the light tread of delicate hooves as an adult roe passed close by me and out into the open. For twenty minutes he grazed quietly, surrounded by a small army of rabbits. Surprised suddenly by some strange sound or scent he lifted his head, pricked his large ears and stood alert. Then he barked gruffly and sharply, turned and, barking every four or five seconds, made his way at a trot back into the forest.

That night I saw no more roe until just before five o'clock in the morning. Then I stalked one until I was within twenty feet of it. It was a young buck feeding near an oatfield on some grassy, heathery hummocks. As I was watching the deer, I saw a rabbit rush past the buck's lowered head. The deer recoiled and the rabbit shot down a hole. A few seconds later the rabbit re-emerged – I think from its size and appearance that it *was* the identical animal – and the roe buck with fantastically slow and exaggerated steps on its thin legs began to stalk the rabbit. Again the rabbit bolted down a hole. Four more times this game was played with evident enjoyment for both animals until the rabbit seemed to tire of the sport or found the new day unwelcoming and went down below for good. For five minutes the roe explored every burrow and then slowly moved away, turning round every few seconds as if to see whether his playmate had reappeared!

At a quarter past five scores of jackdaws – old and young – began to collect in some of the pines and the still summer morning woke to their discordant voices. Then a greenfinch started to sing from the top shoot of a tree and his mate began to give warning to her fledged young of the passage of a wild cat, which was on its way to hunt young rabbits in the grassy clearings. Not far from this spot I once saw a wild cat, handsomely striped with a bushy tail, making its way back to a den under a stony track. Bird song among the pine trees is fading fast in July but one morning a Scottish crossbill, perched on the lead shoot of a tree, began to sing – 'Tip-tip-tip, Tsee-tsee-tsee, Tseeker-tseeker' – a rare event, since sustained songs are not often produced by this bird.

There is a well-grown piece of pine forest near Loch an Eilein – the Loch of the Island – which lies in a high forest of Scots pine among birch-clad hills. The birches ring with willow and wood warbler songs and the pines rise to ninety feet or more over a field layer of heather, cowberry and moss. On the island around which perch and pike swim are the gaunt romantic remains of the castle stronghold of the Cumings and of Alexander, Earl of Buchan – 'the Wolf of Badenoch'. Here the ospreys last nested in 1897 and there is a photograph of the eyrie in volume 1 of T. A. Coward's *The Birds of the British Isles*.

Here I have stood not far from the visitors' centre which provides an introduction to the Cairngorm National Nature Reserve, covering over sixty-four thousand acres and embracing the very heart of the Grampian Mountains; I could hear the echoing calls of cuckoos coming across the loch, the gentle silvery cadences of the willow warblers, the high falling 'Seea-seea-seeas' of tree pipits, and the titter of sandpipers. Beneath these pines were a thick undercover of junipers where goldcrests and hedgesparrows were breeding, and a white glisten from the flowering stars of chickweed wintergreen. Many times I have walked around the loch, finding coal tit nests in hollows at the foot of pines, watching goosanders on the water and sometimes peregrines scything their way overhead. This is a land where once cattle thieves from Lochaber roamed, having left the Spey and made for the forest behind Loch an Eilein and Loch Gamhna; the latter is a lochan covered with water-lilies where I have seen gadwall and listened to redstarts and song thrushes in full voice. There is also another loch where monstrous trout come into the shallow water and with their dorsal fins protruding sweep up with open mouths all the dead and dying insects that have been drifted by the wind into that foot-deep bay.

Summer is rich in birds and the fragrant orchids and asphodels in the swamps. In autumn their place is taken by the beauty of the rowan, birch and the larch. Golden candles of withering birch stand bright against the never-changing deep green of the pines. The old pine forests become more deserted as chaffinches, song thrushes, siskins, woodpigeons and countless warblers leave the woodland. The capercaillies seek the pine trees, where one can still find crossbills, crested tits, coal tits and goldcrests. I have walked through these winter forests and have found long periods may elapse between encounters with small parties of birds. I have watched crossbills feeding in pines near a loch and looking very much like small exotic parrots as they hung and emptied the cones of their seeds or sang short snatches of sub-song to each other. Parties of blackcock would fly up at my approach and sometimes a huge lumbering capercaillie would crash off through the junipers and pines in search of quieter places. I sometimes surprised a few hooded crows, wrens, robins and bullfinches and was even rewarded with ravens flying overhead and engaging passing peregrines or golden eagles that performed fast and amazing aerobatics above the pines.

Many migrant birds passed through Ryvoan in autumn as well, perhaps on their way from the Moray Firth. As I watched the parties of larks, meadow pipits, redwings and fieldfares, swallows, wigeon, greylag geese and whooper swans flying through in the early morning above Loch Morlich I often wondered if I might meet the Red-handed Spectre – Bodach Lamh-dhearg – who according to MacFarlane's *Geographical Collections* 'appears with a red hand in the habit of a Souldier, and challenges men to fight with him' on the lochside, but I never did.

Beyond Ryvoan lie Abernethy Forest and Loch Garten, famous for its breeding ospreys. Pines that grow here are often over 250 years old and they cast their shadow over green hummocks of heather, blaeberry and juniper. These pinewoods too can be rewarding for the observant naturalist and I have seen cow-wheat, chickweed wintergreen, wood sorrel, and lesser twayblade growing in parts of Abernethy Forest. The new reserve of the Royal Society for the Protection of Birds by Loch Garten contains part of the old Caledonian Forest that we also met in Rothiemurchus. I have visited the famous ospreys on a number of occasions but the pinewoods are full of other interesting creatures too. Crested tits churr away in the canopies in summer, crossbills pass by giving their deep 'Tyoop-tyoops' and a gentle chorus of song arises from chaffinches, coal tits and other birds, while capercaillies display in the dawn and fill the forest clearings with their 'tuk-up' notes, explosive pops and wing flaps; once I saw a capercaillie land on the osprey's eyrie, to the consternation of birds and watchers alike, but the crisis was short-lived. Blackcock gather on the moor and fill the air with dove-like 'roo-kooing' notes, plaintive calls and a hissing, exploding 'ker-showw'. I have watched buzzards and eagles on broad wings above the trees and at various times seen sparrowhawks, long-eared and tawny owls, wrens, thrushes, treecreepers, woodpigeons, herons, gulls as well as siskins and redpolls during my early morning walks.

The loch itself is sometimes still, reflecting the pines or the tops of Craiggowrie, Creagan Gorm and Meall a' Bhuachaille, whose reverse I used to see from my camp by Loch Morlich, and sometimes fretted with white wavelets as the wind freshens and the black-headed gulls beat and twist above the choppy brown surface. In still conditions it is possible to hear the lochside resound with the piping calls of oystercatchers and redshanks, the high stuttering of common sandpipers as the male embarks on his bat-like aerial pursuit of the female, the falling sadness of the curlew's liquid song and the softer displays of wigeon, goosander, goldeneye, tufted duck, mallard, teal and gadwall. Red deer often wander through the pinewoods, while red squirrels and roe deer are everywhere. There are badgers and to the south of the loch I visited one sett with many entrances and traces of occupation at one end by foxes. Wild cats are shy and rarely seen. Otters sometimes visit the loch to fish, foxes can be seen making their way along the banks and a winter stoat may appear, bright in his fine white coat.

This is a remarkable world, where I feel very close to nature. Among the birch scrub and scented pines, the forest edge and the more open moor with its grouse, meadow pipits, merlins and perhaps hen harriers I experience a refreshment of spirit and a change of scene that are both unique. Although many people are attracted to the region by the ospreys – and a good number of pairs is now nesting elsewhere but on occasion still being robbed by collectors – and part of the reserve becomes rather populous in the summer, there are parts that remain undisturbed, so that one can in solitude breathe in the special air, listen to the hum of insects and the songs of birds and contemplate the mountains that rise to form a dramatic horizon under an ever-changing sky. In winter and early spring it is a fine thing to look up from the forest of Glenmore towards the angular defile of the Lairhig Ghru, with Ben Macdhui on the eastern side and Braeriach on the western, deep in snow; over the entrance the dark lip of Creag an Lethchoin – Rock of the Half-Dog or Lurcher – lowers menacingly. A great snowfield – Cuithe Crom-the Bent or Crooked Wreath – lies across the face of Cairn Gorm and a white powdering of snow encrusts the conical dome of Carn Eilrig.

I shall never forget my first visit to Loch Morlich nearly thirty years ago. It was on a still summer evening and I approached the loch from just south of Boat of Garten and along An Slugan – the pass through the forest. As I reached the high point I found, spread in front of me, one of the most magnificent natural panoramas in the British Isles. From my point of vantage a green carpet of pines unrolled itself down to the loch which lay before me and below like an oval silver mirror – without a ripple and at peace. In the mirror were sharply reflected the green belts of pines, the blue-brown moorland and above all the snow-splashed summits of Cairn Gorm, Braeriach in the distance and closer at hand Castle Hill where the reindeer roamed, and Airgiod-meall – purple in the evening light. Never again did I see the waters of Loch Morlich so still! There remains a suggestion of the primeval in this world of mountain and pine forest where nature seems dominating and invincible. I have walked for many hours without seeing another human, sharing my days with the mammals and birds. I have recorded their wild voices and filmed their private lives and sometimes I have just listened and looked with a sense of deep privilege.

In August the roe deer of the pine forests mate. This is the smallest and I think the most attractive of our native species of deer – an animal that has probably been living in Britain since the Ice Age. The roe deer is very shy and, since a mature buck weighs only some fifty to seventy pounds, an animal lying up in thick cover may be missed by the most careful observer. Roe deer feed at dawn and dusk and I know several places in Strath Spey where I can guarantee to see some grazing in the evening. The rut is earlier than that in other deer and their antics too are quite different. They form small rings or figures of eight on the ground where the bucks chase the does. I have often come across frayed young conifers showing how the buck has scraped his antlers to mark his territory. The rut is a delightful and remarkable activity to witness. The bucks also bark; their gruff sounds travel easily through a Highland wood at night.

The story of the red deer is rather different. In June the stags have their antlers in velvet and one midsummer day among the old pine trees in Glen Feshie Charles Lagus and I made a film showing the stags standing up on their hind legs and boxing each other with their front hooves so as not to damage their antlers. The red deer come to their rut in October. In August the soft velvety skin on the antlers had shrivelled and stripped off in long leathery shreds. Now the antlers are deep brown and corrugated, with smooth ivory-white points, and the hair on the stags' necks has grown long and bushy. During the October days and nights the roaring, yawning bovine challenges of the stags resound and re-echo through some of the Highland glens.

I knew one region in the foothills – a wild area of heather, bog and scattered trees – where the deer used to come for the annual rut. As the evenings drew in the stags began to take over the mastership of groups of hinds, which they marshalled and guarded as their harems and from which all intruders were driven away. One evening I crawled out with a microphone to wait for a large stag to come to the edge of his territory and challenge his rival a quarter of a mile away. A few minutes after I had settled myself in the heather this animal lifted his head, opened his mouth into the form of a wide trumpet and roared once; then he ran forward twenty yards, halted, raised his head once more and roared again. Then he came close to me – so close, in fact, that I could hear him breathing – and uttered a short, gruff bark. Four times he repeated the bark. Then the night air was riven by a series of shattering roars as he issued his challenges to his rivals – sounds that could be heard a mile away. In this way the stag

205

advertises his position and by repeating his roaring over and over again challenges rivals to dare to come poaching members of his harem. Then suddenly my stag turned and swept back past me to rejoin his hinds on a little heathery knoll. Actual combats are not common, although I have seen some active sparring with the antlers. The rut goes on for about a month but the large stags, which are often the first to acquire harems of hinds, are fatigued after a few weeks and may be unable to defend their harems; they have not eaten perhaps for four weeks. Then a smaller challenger with a greater reserve of stamina may take over the harem, which barely observes its change of ownership and status.

Red deer are social animals but for more than ten months the stags and hinds live in separate herds, the former in loose assemblies and the latter in more closely welded groups. Both sexes observe home ranges and seem to know intimately every tree, rock, ridge and hollow within them. The calves appear from late May to July and are left hidden in the heather for the first few days. If they are mislaid the hinds call to them with low 'maa-ing' bleats. There are reindeer too in the Cairngorms, introduced in 1952; they once inhabited the northern Highlands well into historic times. Near Kincraig is the Highland Wildlife Park, opened in 1972, where the visitor can see ibex, Arctic fox, bear, beaver, bison, lynx, pine marten, polecat, red deer, reindeer, saiga, wild cat, boar, horse and wolf among the mammals, and blackcock, buzzard, capercaillie, golden eagle, peregrine, ptarmigan, red grouse, sea eagle and snowy owl among the birds; the Park tries 'to give you some idea of what we as a nation have lost'.

The Spey Valley seems to have benefited in recent years from an apparent change in climate whereby such northern birds as redwings, fieldfares, pied flycatchers, bluethroats, wrynecks, wood and green sandpipers and goldeneyes have found it attractive. Even goshawks and marsh harriers and bramblings have shown interest and there may be others still to come – perhaps spotted redshanks, great grey shrikes, broad-billed sandpipers and returning red kites. Certainly in the central Highlands I have heard great northern divers calling and visited a loch where they actually nested in 1970.

Not all the surviving pieces of the old pine forest which are of outstanding interest are confined to the eastern Highlands. There is a truly remarkable natural pinewood on the southern shore of Loch Maree in Wester Ross; it is called Coille na Glas Leitire – the Wood on the Grey Slopes. The forest lies in the heart of a fantastic land of rugged mountain, ever-changing sky and cloud, deep glen, ancient pines and heathery moor. The landscape is harsh and magnificent, with great mountain masses such as those of Liathach, Slioch, Beinn Alligin and the seven peaks of Beinn Eighe. Here are some of the world's oldest rocks – hard uncompromising Lewisian gneiss some 1500 million years old, which is overlain in places to depths of two thousand feet by coarse reddish Torridonian sandstone, formed from sediments perhaps under hot desert conditions, and also by greyish-white bands of Cambrian quartzite. Some of the rocks were buckled and folded and then fashioned by ice and river to form the landscape as we see it today. Beinn Eighe gives its name to Britain's first National Nature Reserve, set up in 1951, and whose 10,507 acres, according to the Nature Conservancy Council, embraced examples of the country that make up the deer forests of North-West Scotland – lochside, natural woodlands and conifer plantations, high and low moorland and bare mountain top, lochan, bog and mountain stream.

My wife and I have wandered through this marvellous country-side watching greenshanks, grey and white on long green stilts, goldeneye on a lochan, redwings singing in the woods and around houses, and black-throated divers displaying on the still waters of a loch. All round us hooded crows cawed defiance and willow warblers sang their falling cadences, curlews descended bubbling on quivering pinions, and common sandpipers flitted brown and white along the water's edge. Meadow pipits and wheatears sang above the heath and we have often seen stonechats in their little display flights and been greeted by chaffinches in the lay-bys.

Days and nights of intense and characteristic Highland rain may be succeeded by hours of bright sunlight and shadow. Sometimes the clouds hang dark and menacing over the lower slopes of the hills and damp mist swirls about the glens. Despondent caravaners in ever-growing numbers move, often erratically and selfishly along the single-track roads, in search of better weather. I foresee a day when some kind of control or limitation of access may be needed in these wild parts otherwise the very features that make them attractive in the first place may be destroyed. Some of the single-track roads with passing places help to preserve the character of the Scottish Highlands and the newer faster roads mean more people and, as some of them have little sympathy for wild open spaces, more slaughter as well. I once counted five dead sheep – *sheep*, not lambs of which there were too many to count! – on less than two miles of road. Old black-faced ewes move slowly and deliberately and should present no problem to the motorist, while no one in their right senses would drive between a lamb and its mother. I have also seen stags killed by cars travelling too fast for the countryside in which they live. It would be a pity if easier access led to change and degradation in excess.

For a naturalist like myself one of the gems of the Beinn Eighe Reserve is the fragment of surviving pine forest. I have carried out a number of bird counts in Coille na Glas Leitire and found that the main species 'in order of abundance are willow warbler, chaffinch, wren, robin and tree pipit; another survey gives willow warbler first followed by chaffinch with robin and tree pipit equal third'. There were many other birds living in this pine forest – redstarts, hedge-sparrows, great, blue, coal and long-tailed tits, great spotted woodpeckers, and mistle thrushes. There were no crested tits but I found traces of crossbills in the shape of sheared and shredded pine cones. There is a trail up through the pine forest from the Loch Maree road which passes a burn where otters sometimes fish for sea trout or salmon.

Higher up the slope are stretches of heather, blaeberry and cowberry which are the haunts of fox, wild cat and pine marten. It is in this region that my friend Richard Balharry has carried out his pioneering studies of the attractive marten – the Sweetmart of the Rocks. From this viewpoint I can look across to the blancmange-shaped mass of Slioch formed of Torridonian sandstone resting on the Lewisian gneiss and known as the Spearhead, with its wild goats and occasional eagles, to the patchwork of burnt heather on the opposite slope, to the plantations of pine, larch and spruce where redwings were nesting, and to the cultivated delta of the river's entrance to the loch. Below is the white shape of the Anancaun Field Station where I recorded the pine martens and red deer being studied by Balharry. Red and roe deer browse the young pines; seed is taken to be sown and nurtured for later planting on the lower hillsides. There are bogs among the pines where sundews catch and digest their insect prey. Foxes have their earths and pine martens their dens in cracks and crevices in the rock. There is no finer sight than a marten standing erect and alert at the entrance to its lair with snowy mountains in the background. The lower bogs are the haunts of frogs and newts and many kinds of aquatic insect including beetles, pond skaters and several kinds of dragonfly.

The woodlands above Loch Maree provide a sanctuary and rare haven for wildlife and an experience for the walker that he or she will long remember. Loch Maree is also famous for its salmon and sea trout fishing, I have been privileged to visit the wooded islands, including Isle Maree with its history of pagan rites, St Maelrubha's Chapel and the wishing tree into which hundreds of pennies had been driven edgeways into the bark. To sit on the island in the middle of such breathtaking scenery and to listen to the song of the birds in the pines is as close an approach to perfection as one can achieve in the British Isles.

For many years the Forestry Commission has been planting conifers on moorland on a big scale and we have seen its effects earlier in this book. After planting, the curlews, wheatears, skylarks and meadow pipits have to give way to the warblers, finches and thrushes. The commonest of these planted trees are Norway and Sitka spruce, lodgepole pine, European and Japanese larches, Douglas fir and Corsican and Scots pine. Some of the more mature Scots pine plantations have a wildlife that to some extent recalls that of the old Caledonian Forest. After studying twelve mature Scots pinewoods in northern and eastern Scotland I found that the commonest species of bird were chaffinch, coal tit, goldcrest and wren in that order. There were also siskins, redpolls, crossbills, great spotted woodpeckers, capercaillies, occasional black grouse, long-eared owls, sparrowhawks, mistle thrushes, hedgesparrows, blackbirds, tree pipits, redstarts, spotted flycatchers, woodcock, nightjars and in plantations near the Moray Firth crested tits as well. Where there was any kind of shrub layer I located willow warblers, whitethroats and sometimes chiffchaffs.

These tall plantations with their spires of trees and poor light below are not as attractive to me as the open natural forest. In fact, I had only two thirds the number of bird contacts per hour in the plantations compared to the numbers in the best part of the Caledonian Forest; here the trees are often monumental in growth with magnificent pink-red gnarled trunks and great boughs that push upwards into the rounded bottle-green canopies. These may be the survivors of the original pine forest, much of which is now bare, desolate moorland, stripped by the lumberman and resembling an ancient blasted battlefield with old stumps, dead black boles and scarred heath. Today deep ploughing with great trench diggers has left long parallel ridges that rise up across the contours of the hills and in which the new conifers have been planted – perhaps five thousand to the acre. At first these ridges seem alien and offensive to the eye. The newly planted land is fenced against deer and black-faced sheep and wired against rabbits. Gradually the landscape changes as the trees reach the early and late thicket stages and finally the pole plantation. Areas that were once under trees are re-afforested and views of familiar hills are changed and perhaps finally lost. They may become denser and more varied in their wildlife and are important reservoirs for animals, but we must not forget those fragments of the old Caledonian Forest whose existence was known to Pliny and whose open woodlands can give a clearer idea of what primeval forest was like than perhaps other woodland scenes in Britain. Such a natural asset demands our close concern and since it provides an unbroken link with the Ice Age it requires proper care and conservation.

14. The Summits of the High Hills

In the preceding chapters I have attempted to show how greatly man has directly influenced the countryside by the efforts of his hands and implements and by the effects of his domestic stock. Nothing has been truly natural. Now we come to a region where the land remains clear of his direct influence – the summits of the great mountains in the Scottish Highlands, although some slight pollution may occur here. The Highlands are formed from a great variety of rocks many of which are old in time, tightly folded and squeezed and even altered by such intrusive rocks as granite. The great upland massifs rise with their towering precipices and ice-shaped slopes and plateaux above remnants of the old oak and pine forests and birch scrub, countless lochs and vast often uncharted stretches of moorland. The high summits themselves are generally cold but there are sometimes temperature inversions which can bring about the most dramatic scenic effects. They are often wet and yet in frost they can suffer conditions of severe drought; they are often buried under snow and they may be swept by winds of hurricane force.

Above three thousand feet man has really no part to play and these harsh, arid summits, without grazing animals and clothed in sparse and dwarf vegetation, are in effect untouched, natural habitats. Only when one has walked these hills and their tops, properly clad and shod and prepared for all emergencies, can one begin to have a true understanding of their nature and life at these high altitudes as the weather switches from hot burning sun to blizzard and back again. I have a deep regard for mountains but I am prepared to meet them on their terms. I have never joked about a mountain or treated

211

it with contempt. In this way I have travelled surely but with prudence and in return have enjoyed many rewarding and enlightening experiences. In this chapter I would like to try and share some of them with my readers. I am acquainted with high hills from The Trossachs north to Foinaven and from Beinn Alligin and The Storr east to The Buck and Morven.

For the traveller first entering the Highlands it is probably the Breadalbane Mountains of Perthshire which give the earliest promise of the high mountain world that reaches up to Sutherland. Ben Lawers rising from the north side of Loch Tay to a height of 3,984 feet is the best known – a high massif of largely basic schists above bleak moorland and mat grass growing over glacial drifts. Seen from the road between Carie and Lawers above Loch Tay and its sacred Island of the Blessed Female Saints the lower slopes are grassy with scattered pines and chequered like a chessboard with enclosed pastures, old walls and ditches and cottage ruins; above the summit rises as a shallow cone. One summer morning I walked up through a chain of meadow pipit and skylark song, the caws of crows and the bubble of curlews, while common frogs, which I have seen on many mountain slopes, jumped out of my way on the steady ascent. There was the acrid smell of fox, a stoat was weaving in and out of one of the stone walls and a hare loped off to seek more sheltered feeding grounds. I caught a glimpse of a roe deer but did not see any herds of red. The slope was bright with the winking wings of small ringlet, meadow brown and large heath butterflies.

From brackeny slopes I moved up across a rough pasture of mat grass and through tufts of rush to the summit; near the top a belt of greeny-yellow lady's mantle spread out across my path before giving way to a more rocky terrain with lichens, mosses, fir clubmoss and dwarf willow. The *Scottish Tourist* of 1825 reported that Ben Lawers is 'so easy of ascent as to permit riding to the summit'. Ben Lawers is famous for its alpine flora and its rich and varied plant life is dependent upon the generous supply of minerals with which it is provided. The National Trust for Scotland, which has purchased much of the mountain, has said that this beautiful and unique flora is due to a series of coincidences – 'to a small stratum of ideal rock on mountains which are so high that unstable habitats have existed there continuously since early in the post-Glacial period'.

And what a richness of flowers there is! By the mountain burns are chickweed willowherb, alpine meadowrue and starry saxifrage; on the grasslands mountain pansy and alpine gentian; in the bogs and damp places rare alpine sedges and rushes, bog asphodel and grass of Parnassus; on the rocky slopes and screes mossy and yellow mountain saxifrages, alpine willowherb and scurvy grass, and on the cliffs and some of the ledges alpine saxifrage, roseroot, globe flower, mountain avens and alpine woodsia. The summit is bare and among the stones are fir clubmoss, moss campion and mossy saxifrage but the most varied flora is often in the corries that split the mass of the mountain. For the botanist Ben Lawers is a kind of Mecca and for the ornithologist there are ring ouzels, ptarmigan and perhaps golden plover, buzzards, merlins, and surprisingly swifts that often hunt around the summit itself. Mountain hares are not uncommon and there are reports of wild cats. From the top there are views to the sharp ridge of An Stuc over the shoulder, the distant cone of elegant Schiehallion and Lochan a Chairt lying far below. From the slopes of Meall Garbh Ben Lawers shows up as a truncated pyramid of rock deeply scoured and scooped out on its northern flank. At this height it is often a cold and windy world with extreme temperatures and irregular snowfall. One windy day in autumn I watched meadow pipits on migration, just as I had seen others on Snowdon, in the Lammermuir Hills, along the Cotswolds and over London, and flushed several pairs of ptarmigan on the high plateau. There were snow flurries in the air – a contrast to the warm summer days when Ben Lawers invited one to scramble over the rocks and scree, seeking alpine plants of great scarcity and charm.

To the west lies another mountain in a very different mould, yet one of my most loved hills. Three Scottish lochs are reported to be the most beautiful – Loch Maree, which I described in the last chapter, Loch Awe and Loch Lomond. Loch Awe is the southernmost limit of Lorn and nestles below conical hills at the southern end and under the shadow of the eight mighty tops of Cruachan Ben at the northern end. What a scene this great mountain presides over! The west bank of Loch Awe is largely clothed with conifers and the eastern with broad-leaved trees. Great golden bands of daffodils reach down in spring to the lochside and by the streams blossom primrose, wood sorrel, wood anemone and golden saxifrage.

One week in May I walked amongst the larches just coming into tiny leaf, while the buds on the ashes and oaks were beginning to swell. In the woods of oak, alder and ash near Port Sonachan the bird chorus was growing. A pair of buzzards was nesting in the fork of a lochside sessile oak and I watched the male soaring and banking over the white horses on the wind-swept loch. The common sandpipers had arrived back from Africa. There were oystercatchers trilling on the shingly beaches, hooded crows, mistle thrushes and willow warblers in the copses and from time to time, over loch and brae, came the lovely sound of the curlew's spring bubble. There are

213

many islands in Loch Awe and on one of them – Inishail – The Isle of Rest – the Highlanders buried their dead up to the seventeenth century or so, when packs of wolves roamed Argyll. On a spit of land stands the shell of Kilchurn Castle, built in 1440 by Sir Colin Campbell of Glen Orchy, now a gaunt ruin that has inspired painters and poets as it rises bravely in the shadow of Cruachan. Wordsworth spoke of it as 'silent in thy age, Save when the winds sweep by'. From a nearby alder a tree pipit had a summer singing post, crossbills flew over and swallows and martins were migrating through Glen Orchy with its turbulent rapids and waterfalls, cascades and potholes which have claimed too many human lives in recent years. A whinchat was singing one day from a dead branch in a hillside oak among the grazing sheep – a song utterly in keeping with the wild grandeur of the scenery. A migrant greenshank also called that day across the loch as it passed by.

The countryside between Crinan and Loch Awe is rich in Celtic crosses, carved stones, castles and forts, cairns and circles. At Kilmartin are the gravestones of the Malcolm chiefs, each with an armoured knight in bas-relief, and also the simple stone carving of the crucifixion with Christ's legs hanging straight and the feet crossed. In wooded Glen Shira I visited the ruins of Rob Roy Macgregor's cottage, where the famous outlaw lived for eight years. A steep bank fell sharply down to the river with its pools, falls and strange flat pebbles that glisten like silver in the water; it was here earlier this century that Rob Roy's dirk handle, carved with his initials, was found on a rock ledge by the stream. From the oaks on the brae a redstart was signing, his jingle coming across the Shira as it probably has every year since the end of the Ice Age. On nearby Loch Fyne at Auchindrain is an unusual museum in which old croft houses have been restored to show dwellings from the seventeenth century onwards, with their furniture and implements – a veritable history of life in this part of the Scottish countryside; close by is the spot where the last wolf in Mid Argyll was killed in the eighteenth century by a woman with a spindle, who later died of shock!

Cruachan towers above Lorn, dominating the landscape – 3,689 feet high and some eighteen miles in circumference. Its very name was the rallying cry of the Campbells. I have filmed inside the great underground power station set in the very heart of the granite mountain that stores water in the high corrie reservoir, which can be seen for many miles. To reach the great south corrie I have travelled up the road from the Kirk of St Conan, with its relic of Robert the Bruce – a splinter of bone. From this new track you can reach the summit by the south ridge or cross the dam and aim to the east for the first peak. Often the summit is under snow but then the sun often shines brightly on its peaks around which the golden eagles and buzzards circle and soar; there are red grouse, and ring ouzels and higher still ptarmigan, golden plover and sometimes dunlin. There are fine rewards on a clear day since the view from the summit ridge includes the Paps of Jura, Morven, Rannoch and Ben Lawers. To the north lies Glencoe whose great hills, Bidean nam Beann, Beinn Fhada and Buachaille Etive Mor at the mouth of the glen, formed from rough rhyolite, are a rock climbers' paradise. Here there are dippers, grey wagtails, herons and I have seen black-throated divers. At Carnoch on a steep damp slope is an alder wood full of wrens, chaffinches, willow warblers, robins, song thrushes, great and blue tits, redstarts and buzzards, which breed among the alders and scattered ashes, hollies and beeches beneath which a carpet of bramble, ivy, primrose, violet and wood sorrel unfolds. Ravens croak gloomily in Glencoe where in February 1692 the deep snow in the glen was made red with the blood of the massacred MacDonalds. It is a bleak place, dark with precipices and jagged peaks, where the winds tear through and avalanches thunder down from the high peaks and ridges.

In Glencoe we stand on the threshold of the Central Grampians, that stretch from Rannoch north to the River Dee and from Lochnagar west to Ben Nevis. West of Loch Ericht are the three high plateaux of Aonach Mor, Creag Meagaidh above Laggan and the magnificent bulk of Ben Alder, all over 3,650 feet – marvellous deer country, dominated by tufted hair grass and the home of dotterels. At Drumochter the traveller is in the middle of the wild country of Badenoch – the territory of the Macphersons. The Boar of Badenoch and the Sow of Atholl rise up to the west of the road, separated from Ben Alder by Loch Ericht. To the east are Glas Mheall Mor, Carn na Caim and dark Ghaick of the Seven Corries. To reach the summits of these hills I have slogged through clumps of waist-high heather where emperor moths sometimes fly, flushing many red grouse on the way. After four or five hours of steady climbing I began to surprise ptarmigan and knew that my journey was nearing its end. As I made my way up the last stretch of mountain slope I could hear, borne on the wind, the melancholy whistle of a golden plover. These summits are a strange and remarkable world. It may still be warm and windless at over 3,500 feet or grey and misty, gale-trimmed and swept by torrential rain or whitening blizzard. Woolly hair moss grows up to about 3,300 feet and there are regions of low heath, stony areas and deep scars and channels where erosion has scoured the summits, leaving black hummocks of peat in scattered groups.

Despite even the wind it is an experience to stand under an enormous sky and share this high detached world with the golden plover and dunlin.

One sunny day I left the glens to study dotterels in these western Grampians. This handsome little plover when seen at a distance resembles a small golden plover, but it has a broad white eye-stripe which runs round its deep chocolate head and its upper breast is divided from the lower by a white band. It is a delicate little bird and one of the most confiding of all our British birds – nicknamed the Moss Fool – and this has often led to its undoing at the hands of man. The dotterel is a bird with a strange lifestyle, since the female bird is larger and brighter than the male and takes a dominant part in the courtship. In this matriarchal society, after the eggs have been laid, the male takes on the responsibility of incubation and sits on the eggs for a period of from twenty-five to twenty-eight days; the hen joins other grass widows and seldom visits her mate. One day I watched a male dotterel with four chicks, caught in the same torrential storm that had drenched me and, as the lightning hissed and crackled all round and the thunder crashed and rolled around the hills, the chicks cried out in alarm and were reassured by their male parent.

On another day of perfect calm and sun I climbed up the mountainside, disturbing more ptarmigan, several mountain hares and a party of seven stags. Flies and bumblebees and various clegs flew with me until I reached the top, and rested on a stretch of golden moss in the middle of the dotterel ground. There was still no wind and all was at peace. These high lands were soft to walk upon. The insects departed; there were no mammals, only a few spiders and at first no birds. Two hundred yards away was a dotterel's nest – a scrape in the woolly hair moss with just one egg. For the whole of that day I watched a pair of these rare mountain birds – there are between sixty and eighty breeding pairs in the whole country – as they actually changed position at the nest and gave forth charming little chiming trills and churring notes. The dotterels seem to favour long whale-backed mountains with soft rounded tops and broad summit plateaux and here on the alpine meadows, woolly hair moss and bare stony slopes they nest and hunt for the beetles, small flies, spiders and insect grubs that form their diet. This world of the dotterel can only be properly shared by the men and women able to climb and walk on these great hills of Scotland. On the Sow of Atholl, poised above the Pass of Drumochter, I have been rewarded by views of the flower clusters of the rare purple mountain heath.

East beyond the Forest of Atholl but still south of the River Dee is Lochnagar, while west beyond Loch Treig is Ben Nevis – two imposing mountains at either end of this great central chain across Scotland. Lochnagar rises to a height of just under 3,800 feet – a red granite massif overlooking the River Dee and the Royal Forest of Balmoral with elegant foothills and an awesome crescent of precipices in its eastern corrie that form one of the most dramatic mountain scenes in Britain. Approached from the bridge across the Dee it forms a tremendous climax to the horizon. A broad ridge runs largely above three thousand feet to Glas Meol above the corries of Caenlochan, famous too for its Arctic-alpine plants. On the highest ground there is only woolly hair moss and three-leaved rush but in the corries there are interesting plants like alpine speedwell, highland cudweed, blue sowthistle, brook saxifrage and several rare sedges. I have also seen wood sorrel, rosebay and greater woodrush. Ptarmigan nest on Lochnagar, feeding on shrub heaths – blaeberry and crowberry high up and heather lower down, and some insects in summer as well. This mountain is one of the few haunts of the mountain burnet moth.

At the other end of the chain is Ben Nevis – the highest hill in the British Isles – whose granite summit is often under snow and whose most interesting plants are lower on the mountain slope. Yet more than fifty kinds of invertebrate and the pygmy shrew have been recorded at the summit, where the August average shade temperature is as much as 20° F lower than the average of 57° F in Fort William, inhibiting plant growth, and where an average of 261 gales a year over fifty miles an hour has been recorded. Viewed from Corpach with its great pulp mill and across Loch Linnhe Ben Nevis can be seen to rear its stupendous bulk, but as it may be wreathed or even hidden by cloud for perhaps five/sixths of the year some good fortune in the weather is needed for its proper appreciation and enjoyment. I have approached it from Achintee but the most interesting ascent is from Carn Mor Dearg and then on to the plateau at the top. In winter deep snow cornices hang over the precipices of Ben Nevis and the scenery is then truly breath-taking, but it should be tackled only by experienced mountaineers. In summer the view is tremendous and embraces the Cairngorms, Ben Lawers, Ben Lomond, Cruachan Ben, Jura, Mull and the Cuillins of Skye. I have not heard a snow bunting on Ben Nevis but a pair raised a brood there in 1954.

It is perhaps true to say that summer in Britain may come last of all to the four-thousand-foot mountain tops of the high Cairngorms – a region that I have known and explored for thirty years. My most recent visit was in the summer of 1976 and I was anxious to see these lofty alpine plateaux again where high winds, snow flurries, frost and

217

rain squalls erode the summits and strain the lungs. To reach the tops I travelled up with Richard Balharry, chief warden of the Nature Conservancy Council for the East of Scotland. We made our way through the old Caledonian Forest – dark acres of Scots pine with their crested tits, crossbills and capercaillies and on to open heathery moor – haunts of meadow pipits, skylarks, whinchats and wheatears. That day there was a strong wind blowing and not many pipits or red grouse were prepared to show themselves on the hillside. There were red deer feeding across the valley – we counted several hundred. The mountainside became more steep, the heather shorter and the track more rugged. We paused to study an eagle's eyrie and then finally we reached the high plateau, where we walked on the wind-trimmed carpet of moss and tiny sedge. It was a wide-open demanding world, this mountain land at nearly four thousand feet. Ahead of us stretched deep snow beds, criss-crossed with the tracks of mountain foxes. A few male ptarmigan – those grouse-like birds of the high tops with white wings and underparts – were leaping into the air and descending once more on outspread pinions, croaking a belching 'er-ook-oora'.

Fierce gusts of wind and bitter squalls began to sweep horizontally across the bare plateau. I struggled for breath as we made our way to the shelter of a rocky outcrop and crouched among a pile of boulders high above Glen Einich. Here I recorded a conversation between us for my regular 'Countryside' programme on BBC's Radio 4. It was a remote and very hostile world, in which we took no chances with the terrain or the elements and where our mountain clothes and boots gave us warmth and protection. For many hours we walked on the tops. We saw the first dotterel back from the south, running about in the mist and rain and a goshawk flew by on broad and purposeful wings. All round us we could hear the melancholy and supremely wild and appropriate notes of male golden plovers holding territories – the voices of late-arriving summer in the high Cairngorms.

Geologically the Cairngorms represent a high but rather denuded mass of coarse-grained granite bounded by schists of the Moine series whose montane flora, including mountain avens, cyphel and alpine saxifrage, is scarce or missing from the neighbouring granite. Weather and rivers have between them produced a series of steep valleys like the Lairhig Ghru, which runs from Rothiemurchus to Braemar, wide impressive corries and broad cone-shaped summits of which the highest, at 4,296 feet, is Ben Macdhui where the ghostly spectre of the Brocken is said to walk in the mists. In the high corries of the Cairngorms I have seen alpine lady's mantle, sea pink, starry saxifrage, sibbaldia and moss campion and greater rarities too such as Highland saxifrage and alpine speedwell. Above Glen Einich I have also found among the rocks both purple and yellow mountain saxifrage and alpine sawwort. The corrie lips that lead to the different summits are sometimes covered in sedges and rushes, while the tops themselves have only a sparse vegetation of lichens, moss, three-leaved rush, spiked and curved woodrush, sedge, viviparous fescue grass, dwarf willow, starry saxifrage and moss campion. Late in June the Cairngorms plateau is sometimes pink with the tiny scented flowers of moss campion, to which black burnet moths will come.

There are several ways to reach the high plateau of the Cairngorms. One day many years ago I scaled the cliff of Creag nan Gall above the Green Lochan in Ryvoan and made my way to Mam Suim, past Stac na h-Iolaire – Eagle's Crag – and then below the rock outcrops above Coire Laogh Mor to the top of Cairn Gorm; now, of course, I could walk up from the ski-lift at the top of the mountain road. As I reached the summit I came across small parties of ptarmigan which stopped to watch me, croaking all the time; most were confiding – this was before the advent of the ski-lift – but some dropped down into Coire Raibert. Blue hares loped away from me in their curious dog-like fashion and several deer hinds allowed me to approach them within sixty yards. A golden eagle was soaring with consummate ease over a dark and menacing precipice and I could hear ravens croaking in the distance. The air was very clear and almost without wind, and I made a tape-recording at just below four thousand feet with no sounds of wind blowing through the dwarf vegetation, no calls of birds, nothing! I have also picked up small crystals of Cairn Gorm stone – sherry-coloured lumps of quartz – but nothing remotely like the stone weighing fifty pounds that was sold to Queen Victoria.

On other occasions I have climbed up past Jean's Hut to Cairn Gorm, negotiated the Lairhig Ghru to reach the central plateau, advanced up Glen Einich or Glen Dee. Yet one of my favourite approaches has been from Achlean in Glen Feshie, where I used to stay with the late John Clark. Then I would walk up the lower slopes of Carn Ban Mor, past the pinewood by Allt Fhearnagan, and follow the stalker's path and its famous zigzag. I always stooped to drink at the burn draining out of the great snowbed of Ciste Mearad – Margaret's Coffin; it is said that Margaret, having been jilted by Mackintosh of Moy, died here in her mad wanderings. The snow-bed is roughly heart-shaped, lying in a depression below the summit. The water from it – cold and pure – is a delight to drink. In 1957 I lay by the patch of snow recording ptarmigan for the first time in

219

Britain, since a small recorder had only just become available. One day I was burned by the fierce sun and the next frozen by wind and rain.

When you move up and reach the summit of Carn Ban Mor and the world of golden plover and dunlin a vast, unexpected plateau unfolds before you. From here I was able to investigate many square miles of the mountains. I followed the wave-like crest of Sgor Gaoithe, where I could look down through a tumult of rock to Loch Einich with its Arctic char, and from which I could see across the glen to the even more precipitous flanks and slopes of Braeriach. Many times I watched red deer moving in a thin grey-brown line below the waterfalls above Coire Odhar. I have walked up to the summit of Sgoran Dubh Mor but, as it stands back from the edge of the rocks, it does not provide so fine a view into the glen. On other days I have been drawn towards Am Moine Mhor – The Great Moss – a peaty corrugated region of some difficulty where golden plover call and dunlin trill and whose little lochan, Loch nan Cnapan, sometimes holds common sandpipers, even dippers on occasion and scores of black-headed gulls. In winter there are snow buntings here in small brown and white swirling flocks, and one spring I had a brief glimpse on the River Eidart of a bird that I feel sure was a spotted redshank. In the other direction the plateau leads to the great swelling slopes of Braeriach, the second highest mountain in the Cairngorms and the third in this country, standing at 4,248 feet, and to the rounded shape of Monadh Mor and the square barn-like top of Carn Toul. From the Great Moss there is a direct route above Coire Dhondail, which drops dramatically down to Loch Einich, up to the Wells of Dee and thence to the summit of Braeriach. The River Dee itself falls from Braeriach into An Garbh Coire Mor and according to James Hogg over

> the grisly cliffs which guard
> The infant rills of Highland Dee.

The Dee cascades past snowbeds which lie like white islands in a vertical dark sea and some of which rarely disappear. At their thawing edge flourish plants that can survive a ten months' snowfall every year – starry saxifrage, dwarf cudweed and various kinds of moss. Hinds often feed up to this height in summer and mountain hares breed near the Wells of Dee. Until early May snow buntings occur on this plateau and sometimes a cock bird – starkly and magnificently black and white – takes up a territory in a corrie; here he calls 'tseep' or sings a fluting, lark-like stanza in a little song flight and sometimes trips charmingly across a snow bed. The nest is often tucked inside a mass of boulders and difficult to find. There are two races of snow bunting involved in Scotland – one from Iceland and the other from Greenland and Scandinavia, with the latter having more white on the rump. To see and hear snow buntings on the roof of Scotland as I have done is an experience almost impossible to match for the birdwatcher, and it has taken place in a natural rock garden of moss campion and creeping azalea. Ptarmigan slip away with their nestlings on the slopes of Beinn Brodhain above Glen Geusachan; here one can still see the ancient roots of pines that gave their name to the Glen of the Pine Forest, where perhaps the jet-black demon hound Brodhain still stalks. Another of my walks has been across to Meall Dubhag to look down into Coire Garblach and Glen Feshie; there is a scramble down the scree to the shingly river with its noisy oystercatchers and fluting ring ouzels, and the ancient pinewoods with their chaffinches and coal tits.

There are many happy memories for me: of Braeriach in the moon-light, snowy and Olympus-like; of the sheer winter grandeur of a snow-filled Lairhig Ghru; of the craggy scarred sidings of Sgoran Dubh Mor; of the panorama of summit ridges, slopes and peaks as seen from Ben Macdhui and running as a line from the Devil's Point across Carn Toul, Braeriach to the Pools of Dee and the V-shaped nick of the Lairhig Ghru; of deeply sunk Loch A'an – remote and stern, of which Queen Victoria wrote 'nothing could be grander or wilder', and which lies under Shelter Stone Crag and its protective boulder – Clach Dhirn; of Loch Etchachan beneath the tops of Ben Macdhui and Cairn Gorm with its emaciated trout at perhaps their highest point in the British Isles; and of swifts hawking for insects above the golden plover on the Great Moss.

One of my most vivid recollections is that of a descent from Carn Ban Mor down to Glen Feshie. I had a huge pack of tape-recording equipment on my back and was coming down slowly, leaning for-ward with my shoulders hunched, presenting a very non-human kind of figure to the world. Much of the time I had been looking down at the footholds on the ground but I suddenly became aware of some figures moving to the left on the very periphery of my vision. I slightly inclined my head to discover that a party of stags was moving alongside at the same speed. I kept on walking and soon two or three of the animals had come within twenty feet. For minutes on end the stags descended with me – curious and quite unafraid of this unfamiliar shape. Then I paused for a split second to adjust a strap on the pack which was starting to bite into my shoulder and they were off like the wind. It might have been long enough for me to sight a rifle and they kept running until they had quite disappeared over the distant brae.

In my search for the countryside and wildlife of Britain I have visited many of the less well known mountain ranges to the north. I have watched golden eagles tumbling and diving above Glen Shiel, set against the natural backcloth of the bristling Cuillin peaks beyond Loch Alsh. I have explored the hills of Torridon in Wester Ross. There is Beinn Eighe with its sterile quartzite slabs and pavements – the so-called File with its ridge of seven peaks seven miles long – much of it red sandstone with some peaks entirely of Cambrian quartzite. I have seen flowering dwarf cornel, but the quartzite tops are bare and impoverished compared to that of Ben Lawers.

There is a National Nature Reserve at Beinn Eighe and there are ravens, peregrines and golden eagles and many red deer. Like nearby Liathach the File is formed from Torridonian sediments capped later by the sterile Cambrian quartzite, which gives an appearance from a distance of an icy, even snowy coating to the mountain. Its great glory is one of the most magnificent corries in the Highlands – Coire Mhic Fhearchair – where three precipitous terraced cliffs rise like threatening fangs from the screes and lochan at the base. In winter the snow lies along the parallel ridges of sandstone and the lower part of the cones of Beinn Eighe, while the quartzite peaks above rise smooth and voluptuous like a white breast. On the high summits there are ptarmigan, golden plover and red deer, whose voices can be heard against the ghostly sound of distant waterfalls and in a damp air of deep mists which swirl around the dwarf juniper, sea pink, tufted hair grass and great woodrush, clinging to the barren surface where perhaps one might just find a snow bunting or dotterel.

This is superb mountain country, with Beinn Alligin – the Jewel Mountain – with its 'horns' or cluster of four snow-topped peaks above Loch Torridon, Beinn Dearg with its rosy 'turrets' and Liathach – for me the most imposing, massive mountain in the British Isles. From its summit a long view stretches north across the lower hills to the An Teallach range, with its array of Torridonian peaks, its forbidding Toll an Lochain corrie and dizzy summit crest above Dundonnell Forest. The slopes of An Teallach – named The Forge from the smoke-like mists that I have often seen swirling around it – rises out of blanket bog, and its slopes are stony or overlaid in places with mat grass, tufted hair grass and occasional three-leaved rush. There are sometimes eagles, peregrines and ravens along its jagged mass.

Still further to the north, where for me the wild Highlands really begin, are some of the finest and most exciting scenic views in

Britain. One June day some years ago produced an absolute kaleidoscope of colour – the tropical gardens of Loch Ewe, the rhododendrons and broom in extravagant abandon along the roads, the quite heavenly blues of Gruinard Bay and Loch Broom, and the white houses huddled together in Ullapool. There were the more sombre greys, whites and blacks of the hills – the superb long ridge of Ben More Coigach approached through the birch scrub at its base, cloudy Quinag shaped like a Y with its five tops and two main peaks – sandstone again resting on the gneiss, the low flat triangle of Canisp, weather-shattered Stac Polly like a fairy castle with battlements, Cul Mor – twin-summited with its sharp precipices and many red deer, striated Cul Beag, and Ben More Assynt, the highest mountain in Sutherland. There is a very special quality about this land of hill, loch and fjord. Of all the mountain ascents I think that the climb up Stac Polly above Loch Lurgain is one of the most rewarding. Its crest consists of shattered Torridonian sandstone pinnacles, which with the loss of the previous quartzite capping have become eroded into a fantastic forest of stone pillars. From the top the view is almost indescribable. On one clear day I looked across a veritable wilderness of moors, flows and blue lochans dominated by the peaks of Cul Beag and Cul Mor, the ridge of Suilven and the triangle of Canisp and I wondered what rare plants and animals, perhaps undiscovered ones, lived in this remote and unapproachable countryside. As I came down to visit some of this country I came across red and black-throated divers and greylag geese, with their wails, quacks and honks, and heard the plaintive calls of greenshanks, while willow warblers and redwings were singing in the scrub below the mountain.

I am also greatly attracted to Suilven – the Pillar or Sugar Loaf – which to the inhabitants of Lochinver presents a dramatically steep cone but from Stac Polly appears as a long ridge like a resting lion. It is an impressive sandstone hill rising up from a riot of bogs and lochans on the Lewisian gneiss and with great walls falling from Caisteal Liath – the Grey Castle – which is its summit. It has been compared to the Sugar Loaf at Rio de Janeiro, and viewed from the sea as I have seen it the comparison is a fair one. A difficult climb to the summit ridge which runs for a mile and a half reveals the shapely sharp pyramid of its secondary summit – Meall Mheadonach – and of the more rounded Caisteal Liath. The ridge is made up of shattered sandstone rocks and white quartzite boulders amongst which woolly hair moss and a little grass manage to grow. To the east there are views to Inchnadamph Forest and Ben More Assynt – 3,273 feet of Lewisian gneiss resting on Cambrian quartzite. As a result of geological movements areas of limestone have come to the surface; those especially at Inchnadamph, Knockan, Elphin and Durness are home for many fine mountain flowers which often grow at quite low levels.

In this region the botanist can find mountain avens, hoary whitlow grass and dark red helleborine orchid, all of which we have met before on the limestone, but there is also the rare Arctic or Norwegian sandwort. The green limestone country, with its grass, farmsteads and calling corncrakes and few jingling corn buntings, lies clearly separated by the eye from the adjoining dark brown peaty area on the acid soils. A walk up the Traligill Burn from Inchnadamph on to the slopes of Ben More Assynt is a rewarding one, and I have seen globeflowers and the rare Don's twitch, while on other sites I have listed purple saxifrage and yellow mountain saxifrage. In wet weather it is interesting to watch the streams and waterfalls gush out of the face of what is normally a long dry cliff. The limestone country reveals karst pavements, sink holes, caverns and underground streams, as one might expect. By Allt nan Uamh is a cave in which the bones were found of northern lynx, bear, Arctic fox, reindeer and human remains attributed to the period around 6000 BC. Red deer come to one favoured spot I know near Inchnadamph and the neighbouring Inverpolly Reserve, which includes the Knockan cliff, lochans, birch-hazel woodland and several peaks, among them Stac Polly, provides habitats for golden eagles, red and roe deer, wild cats and pine martens.

To the north again is Reay Forest, once called Direadh Mor – the Great Ascent – and here is another wild piece of country of lochans and moors and strange stark hills, the barest in Scotland, which rise to the sky and enclose vast tracts of land. There are the great precipices and rocky shoulders of Arcuil – Ben Arkle – part of a great horseshoe of uncompromising white quartzite banded with white scree – a unique mass where grouse feed and ptarmigan browse down to as low as fifteen hundred feet. The name Arkle is best remembered for its association with a famous steeplechaser, but Sir Robert Gordon noted in the seventeenth century that 'there is a hill called Arkill; all the deer that are bred therein, or hant within the bounds of that hill, have forked tails, threi inches long'. Alas, these strange animals are no more!

The great mountain of Foinne Bheinn – Foinavon – which again has a name associated with a famous horse – is remote and difficult to reach, rising as it does from a huge moor and flowland of lochans where golden plover, dunlin and divers breed. Foinavon – the White Mountain – dominates Sutherland in a very special way and its high winding ridge reminds me of Striding Edge on Helvellyn. There is a mystic whiteness about the mountain. A Cheir Gorm is a striated ridge or promontory lying on a sloping bed of white scree. The mountain is bare but from A Cheir Gorm it is possible to listen to the shattering fall of quartzite blocks around you. Perhaps snow buntings may nest here from time to time. The ridge gives a view of the other mountains of Sutherland – Ben Stack – a cone rising up beside the loch famous for its salmon and sea trout, Ben Hope to the east – the most northerly mountain over three thousand feet in Scotland, which guards another land of flow bogs where greenshank breed, of lochs and the Pictish broch of Dun Dornaigil, Ben Loyal prominent beyond – the so-called 'Queen of Scottish Peaks', and to the south-east the humps of Ben Hee, Ben Klibreck and Ben Armine. In recent years dotterels have been seen in summer on Ben Loyal and Ben Hope without being proved to have nested and there is a dotterel in the Dunrobin Museum in Golspie labelled 'Ben Klibreck, 18 June 1846'.

Before I close this chapter I should make at least a brief reference to the Isle of Skye, since I have visited the Cuillins and climbed the Storr on the promontory of Trotternish. The Black Cuillin and Red Cuillin – the Keel-shaped Ridges – reveal frost-shattered peaks pushing out into knife-like ridges separating corries with vertical cliff faces. The gabbro of the range has weathered better than the basalt and the seven-mile horseshoe provides some of the safest and best mountaineering in the British Isles. Peaks rise to more than three thousand feet but few can be reached without rock climbing. Near Sligachan, where I have heard greenshanks, is Sgurr nan Gillean – the most northerly peak of the Cuillin, while Glen Sligachan gives access to Loch Scavaig and the main Cuillin ridge, or to Loch Coruisk.

The Storr has fine basaltic cliffs and pinnacles including the tall stack known as The Old Man of Storr, which I have reached from Loch Leathan. Here there is green turf where mountain hares disport themselves and such plants grow as eyebright, northern rock cress, Arctic sandwort, moss campion, cyphel, saxifrages, lady's mantle, alpine sawwort, roseroot and holly fern, and that special glacial relict species Iceland purslane, first discovered in 1934. My route up to this floral haven is by means of the north-east flank which overlooks the road from Portree to Staffin Bay. Viewed from a distance it is possible to see that the Storr is a tilted sloping mass rising from the west that ends in sharp cliffs and pinnacles in the east. From the top there is on a clear day a sight of Clisham in the Isle of Harris, views across the Sound of Raasay to Applecross – now

225

opened up to the world – and the Bealach na Ba – a road at 2,053 feet where the ptarmigan call as you drive up the hairpins above a spectacular landscape, to Slioch above Loch Maree and the hills of Torridon. I have sat up on this plateau enjoying a tremendous wealth of visual experience. This was once the home of the white-tailed eagle but there are still golden eagles, peregrines, buzzards, ravens and ptarmigan. Skye has almost everything.

Why do the high summits of the mountains have such an appeal for me? Firstly, they have to be reached by physical effort and application, and an ever-widening view is a constant reward for the determination. Secondly, there is acquaintance with some of our rarest birds and alpine flowers, set often in outstanding scenery. Thirdly, there is that unique world of moss, lichen, starry alpines, eroded banks of peat and the glitter of bright pebbles where human life seems to be standing still. The feelings that I experience as a naturalist mingle with those I enjoy as a man and for a few brief hours I am refreshed and rehabilitated. Certainly my feelings on these mountain tops cannot be matched by those I undergo in any other kind of habitat. Then there is the pleasure of the descent into the more mundane world whose compensations are the fertile lowlands, luxuriant green leaves which contrast with the greys, blacks and purples of the high tops, and a whole new spectrum of flowering colours.

15. Islands in Grey Seas

There is a very special magic about islands. Many of us dream of retirement to some tropic isle or some sea-girt reef off our own shores in the British Isles. The very word 'island' elicits romantic feelings inside us, and the sea? – well, in the words of Eugene Lee-Hamilton,

> It is the blood
> In our own veins, impetuous and near.

I have already written in this book of visits to some islands – Anglesey, Skokholm, Skomer and Grassholm off the coast of Wales, of the Isles of Man, Wight and Skye – the Isle of Mist, of Rathlin, Cape Clear and the Aran Islands in Ireland, and of Holy Island and the Farnes off the coast of Northumbria. I have been fortunate too to see Ailsa Craig with its gannets and Arran in the Firth of Clyde, Islay – green and peaty with its distilleries and Greenland barnacle and white-fronted geese, Jura with its conical quartzite Paps rising from a waste of blanket bog and deer ground, Tiree with its flowers carpeting the ancient gneiss, Colonsay and Oronsay of Torridonian sandstone with their moors, scrub and ancient remains, volcanic Mull with its high peak of Ben Mor, ancient castles and amalgam of mountain and moor, and, of course, Rhum. The last is formed from sandstone and gabbro and boasts fine cliff scenery, experimentally managed red deer herds – it is a National Nature reserve – wild goats and ponies, Manx shearwaters and recently re-introduced white-tailed eagles. I have sailed between the islands of Canna – the 'garden of the Hebrides' – and the tiny basaltic island of Sanday, and cruised around the Summer Isles off Achiltibuie with their mergansers, black guillemots and seals. I have seen the splendidly magnificent Shiant Isles in the North Minch with their 500-foot columnar basalt cliff; these are Eileanan Seunta – the 'Enchanted Isles' – once owned by Sir Compton Mackenzie and the home of perhaps some eighty thousand pairs of puffins as well as many fulmars and other seabirds and both black and brown rats. Island visiting has been one of my great pleasures as a naturalist. There are some islands or island groups which are more varied or more remote than others, with contrasting terrain and habitats that have had a special appeal for me and some of these I have chosen for this present and last chapter in the book.

At the extreme south-west of the British Isles are the Isles of Scilly, lying some twenty-eight miles west-south-west of Land's End in Cornwall. There are five inhabited islands – six, if one counts the Gugh, which is attached to St Agnes by a sand bar – and these include the largest, St Mary's, and St Martin's, St Agnes, Tresco and Bryher. There are also some 140 other uninhabited islets and rocky fangs of granite. It is not given to many of us to set the season back at will but this I was able to do one December. I visited the Fortunate Islands, as they are called, not by the *Scillonian* from Cornwall but by the helicopter from Penzance heliport, which provided me with a fine bird's eye view of Newlyn, Mousehole and Porthcurno on the mainland and both the Longships and the Wolf Rock lighthouses. The amphibious Sikorsky helicopter put down on a tarmac pad on St Mary's. Here on a grass sward of the freshest green were a dozen fieldfares and large numbers of redwings. In a hired van I rattled down from the heliport to the island's capital – Hugh Town – which lies astride the narrow isthmus, dividing the main part of the island from the Garrison and Star Castle, built in eighteen months between June 1593 and December 1594, to resist the Spaniards after the Armada. Around the Garrison I felt that I was back three months, perhaps more, in time. I could see daisies and dandelions in flower as well as ragwort, periwinkle, fumitory, red campion and vast amounts of Bermuda buttercup, while the gardens blossomed with fuchsias, carnations, campanulas and nasturtiums. There were numbers of small warblers in the hedgerows, while four house martins and a swallow, detained by the attractions of the mild climate, hawked for insects above the beach. Two black redstarts were flitting and bobbing among the bits of ripe stranded seaweed together with a couple of stonechats. There were paeans of music from song thrushes in the trees near Star Castle and even in Main Street, Hugh Town.

Later that morning the *Scillonian*, with a mixed company of attendant gulls, berthed in Hugh Town, bringing in cars, carpets, meat, hardboard, drums of fuel, clean laundry and kippers –

reminders of the island's isolation. On the opposite side to the quay on a sandy beach turnstones, ringed plovers and tiny delicate white sanderlings were running back and forth in front of the tide. The trilling of the turnstones and the twitter of the sanderlings were perhaps the commonest natural December sounds on St Mary's. Starlings, house sparrows, rock pipits, song thrushes and even robins foraged for food on the beach. In Old Town Bay, below the palm-fringed churchyard with its many victims of Scilly's past wrecks and below a veritable living carpet of mesembryanthemum flowers, was a sandy, muddy strand purple with the bloated corpses of stranded jellyfish. On these beds of ooze I could see curlew, ringed plover, oystercatchers, redshanks and even two greenshank probing and picking up morsels of food.

For the people of Scilly life has always been governed by the sea. Now they grow flowers, run pleasure boats and accommodate visitors but the last two activities put new pressures on the islands' habitats, the sea urchins and the rest of the wildlife and cause some disturbance. Special permission is necessary before landing on the uninhabited islands and this gives some protection to the seals and seabirds. The Isles of Scilly are parts of a submerged granite mass and enjoy a warm climate where snow and frost are uncommon. The sea is ever-present – a reminder of the Scillonian sea captains, who, after voyaging round the world, came back with new and strange plants. The mainstay of the present flower industry is a small, branching and sweet-scented golden narcissus of great charm – the *Soleil d'Or*. The 'Sols', as they are called, the Scilly and Paper Whites and daffodils such as King Alfred and Fortune, grow on little box-like plots sheltered from the south-west winds by tall 'fences' – in Scilly that means hedges – of pittosporum, escallonia and veronica, which were often themselves encompassed by shelter belts of pine. As I walked round the plots I found that the hedges were alive with the noisy chatter of redwings and the clack of blackbirds. Two miles out from St Mary's and to the east lies St Martin's, with its white beaches and gentler scenery. Westwards is Tresco, with its caves, museum of ancient ships' figureheads and sub-tropical gardens, where I heard a chiffchaff singing softly in December. There is Bryher, known as 'the pearl of Scilly', twin-humped uninhabited Samson, where I saw the gulls gather in long white lines, and remote St Agnes, wild and jagged on its western seaboard, with its Bird Observatory, sand bar to the Gugh and swift dangerous tides. Boats enable one to see something of the wildlife of the islands and reefs, each with its own attraction and charm. I can recall mid-winter days when the sea roared and crashed in on the granite rocks, casting white spray over the weathered piles of granite – smooth shapes built up like giant currant buns. On St Mary's the Hercynian granite shows both vertical and horizontal joints due to weathering, and is formed into blocky 'tors' like those we met on Dartmoor.

In summer goldfinches and linnets add charm to the rather colourless bulb fields and stonechats 'tick' on the miniature moorlands. On Tresco sedge warblers chatter and whistle away in the reedbeds on the Great Pool and thrushes, blackbirds and wrens scold and titter in the shrubberies, while some of the thrushes search for sandhoppers on the beach. Rock pipits parachute down over the shore and oystercatchers pipe. Ringed plovers also can be heard muttering away on the beach. There are rabbits on Scilly and Scilly shrews, mice, bats and colonies of grey seals especially on the western rocks. The islands too are famous as a breeding place for seabirds. Some of the small rocky outcrops clothed in bracken, brambles, sea pink and Cornish mallow are graced with such mellifluous names as Mincarlo, Menavaur, Rosevean, Rosevear, Menawethan, Gorregan and Great Ganilly. They are unapproachable except in calm water and require landing permits in the breeding season. Annet is the largest and most famous of these seabird islands. Razorbills, guillemots, puffins and shag breed and there are colonies of kittiwakes, fulmars, storm petrels and Manx shearwaters which nest in burrows under the springy turf; at night the shearwaters punctate the darkness with their ghostly wails. Many rare bird migrants appear in spring and autumn as well. In 1977 twenty-nine different species of rare bird were reported in the Isles of Scilly, including white stork, Wilson's phalarope, olive-backed pipit, alpine accentor, Radde's warbler, Spanish sparrow, black and white warbler and rustic bunting. As one might expect, a fair proportion of the records are of birds from the New World.

Due north of the Isles of Scilly by some 550 miles are the 'Isles on the Edge of the Sea'. Of these the Outer Hebrides, beyond the wide channel of the Minch to the west of the mainland of Scotland, are the largest. Their 130 miles of archipelago form what is often called The Long Isle. The special character of these islands has also, in part, been determined by the nature of their rocks – pre-Cambrian Lewisian gneiss; this is the oldest rock of all in the British Isles, formed from a coarsely banded metamorphic material. The landscape is generally one of low rounded hills with some granite outcrops – grey ice-smoothed humps rising out of black peat moorland. I have visited the Isles of Lewis and Harris several times – by boat and by air. I have flown to Stornoway – it has been said that the name is like a banner! – past the great cliff at Eilean an Tighe and

230

over the brown and grey Isle of Lewis. At Stornoway there are trees and a softer framework than for most island towns. Stunted willows and sycamores grow along the roadside dykes and in the grounds of Lews Castle great banks of rhododendrons run in wild profusion among the stands of mature trees. I have been there in winter when a bitterly cold wind of gale force was blowing and large flocks of rooks and jackdaws were beating across the harbour and calling low over the sea against the roar of wind and tide. Indeed my first experience of the island was in deep winter and I travelled west to see the black houses and to look at what is regarded as the most important Bronze Age monument after Stonehenge – the famous circle and cross of the Standing Stones of Callanish, known to Greek scholars as 'The Great Winged Temple of the Northern Isles'. Not far away is the broch of Dun Carloway – the best preserved monument of its kind in the Hebrides – with its starlings and fine views out towards the Atlantic. The eastern flank of Lewis and Harris is grim and rocky, culminating in the north in the Butt of Lewis where the isle points towards remote North Rona and where the sea pounds the rocks around the lighthouse. I have many times stood by the lighthouse looking north-east out over the grey sea. In winter wild geese pass over in small numbers on their way to the west coast and snow buntings dance white and brown across the rocky cliffs, their calls penetrating the sound of the surf. One November I feel sure that I heard and saw a Pechora pipit from Siberia.

The West coast is fringed with shell-sand – silvery white banks of multi-coloured fragments – although rather sparsely so in Lewis. Here there is a strip of calcareous 'machair' – sweet grassland on which are reared large numbers of cross-bred cattle. I have seen the 'machair' on islands from Coll with its summer Arctic skuas north to Lewis. This natural grassland burgeons with flowers in summer – clovers, vetches, storksbill, cranesbill, thyme, harebell, primrose, buttercup, daisy, thistle and many kinds of grass over which great clouds of common blue butterflies dance and flutter. On the ground large numbers of beetles, harvestmen and spiders live out their lives. The central part of Lewis lies under a blanket of black moors, peat hags on glacial boulder clay and myriads of freshwater lochs where red and black-throated divers nest. It is land that man cannot inhabit permanently but offers peat for his fuel. The townships nearly all lie on the mile-wide strip that borders the sea coast and their names – Callanish, Carloway, Arnol, Braga, Barvas and Brue – have a strong Norse ring about them. In the township of Brue, which I filmed for a television programme, the crofts are from six to seven acres in size and slope gently down in their narrow strips towards the sand-fringed loch. In November some were already planted with oats and potatoes and one supported an Ayrshire cow and her calf. There were five thousand acres of common grazing on the nearby hill and I saw the Lewismen claiming some of their black moorland for grazing by spreading ten tons of shell-sand to the acre and following this with five hundredweights of phosphates and three hundredweights of compound fertilizers. Then clover and fescue were sown and in a short time sheep were able to graze the new growth.

A Lewis crofter spent perhaps ten per cent of his time on his croft; the rest was passed in making Harris tweed, which to qualify for the Orb trademark had to be woven in a crofter's home in the Outer Isles. Work is hard but the Lewisman is versatile and a master of many techniques and skills. He is farmer, weaver, carpenter, engineer, boat-builder and fisherman, mason and house-builder. His island can be swept by grey gales or lie warm under a benevolent sun, but his whole way of life is shaped by his island home and the natural history of the Western Isles reflects the isolated nature of the region. The land – and seascapes are often almost beyond belief.

I have travelled every road in winter and summer on Lewis and Harris from the Butt down to Rodil at the southern end of Harris with its church of St Clements – the most outstanding ecclesiastical structure in the Western Isles, looking out across the Sound of Harris to North Uist. There are broad sweeping bays with waters of purple, royal blue, emerald or angry black and great crescents of untrodden white sand. Here ringed plover, godwits and redshank assemble in great winter flocks and one of my most vivid memories of this beautiful west coast of the Hebrides is that of curlew flying and calling above these empty bays. Another memory is that of the Thrushel Stone, which stands near Barvas like a sentinel above the sea – a magnificent menhir rising some nineteen feet above the ground, surrounded appropriately by scores of fieldfares, dozens of redwings and smaller numbers of blackbirds flying low over the moorland.

In summer Lewis and Harris undergo a transformation. On the moors wisps of smoke rising from countless fires betray the men and women busily raising the peats and there is activity on the shielings and around the tiny huts dotted over the black-brown moor. Sea pinks and other flowers cast a delicate rosy glow over the cliffs where the shags and auks are breeding. Fulmars cackle, kittiwakes yell and auks caw and moan above the sea, divers honk around the lochs, skuas wail and skylarks and blackbirds sing above the crofts or on the houses and barns. As in other parts of the Outer Hebrides the commonest sound is the toothcomb rattle of the corncrake on the crofts. Lews Castle is full of bird song and in the woods I have heard ravens, rooks, crows, starlings, woodpigeons, collared doves, blackbirds, song thrushes, robins, blue tits, treecreepers, white-throats, willow warblers, chiffchaffs, goldcrests, grey wagtails and tree sparrows, while the Hebridean wren's song, which I was the first to tape-record, with its attractive trills is perhaps the most distinctive of all wrens' songs. Summer means days of intense blue sky and hot sun, spring squills, primroses, acres of white daisies and dandelions, yellow iris and kingcups and roadside dumps of tweed awaiting collection. Harris is divided from Lewis by a chain of high hills, with eagles, ravens and red deer that have unusually small heads. Their peak is Clisham at over 2,600 feet, which looks down on ice-worn hummocky, tarn-dotted scenery and fjord-like submerged valleys. One of my favourite journeys in summer is along the road from Tarbert to Husinish – the longest thirteen miles in the Outer Hebrides. Here I have seen sunsets of remarkable glory and also St Kilda, silhouetted against the distant sky – a tiny triangle of land some forty-five miles away 'in the lap of wild ocean'. Nearby I have seen rock doves, fulmars, shags and other seabirds at home. Harris is noteworthy for its lazybeds – narrow oblong platforms of peat set in shallow rocky hollows, covered in seaweed and then planted with oats and potatoes. I have watched starlings and wagtails on these lazybeds, agricultural enterprises which I have always regarded with no little awe. Some of the best can be seen near Beacravik. Seaweed is collected for other uses and the product in the form of alginates is obtained in a special factory on Loch Erisort.

The Long Isle stretches south to North Uist, Benbecula and South Uist. All these islands have been linked since the Second World War by a road whose causeway crosses over the old tidal fords. Once one has reached Lochmaddy by boat from Skye or Harris then a car will take you across all three islands but the journey will not include all the centres of village life or places of interest to the naturalist. After Lewis and Harris North Uist seems incredibly green and fresh. It endures mighty gales in winter but by April the high winds, rain and snow flurries may start to subside, giving way to days of calm when the geese can be heard going over and the corn buntings singing their metallic jangling notes. North Uist I find full of interesting sound – the piping of waders, the screams of gulls and terns, the laughter of shelduck, the low quack of gadwall and the cooing of scaup. There are acres of flower-studded machair and marshes full of kingcups and bog bean where snipe, redshank and lapwing call. There are cuckoos, short-eared owls, kestrels and once I saw a merlin.

Balranald on the west coast of the island is an RSPB reserve-home of corncrakes and corn buntings, and I always associate the site with that delightful and increasingly rare wader – the red-necked phalarope. I have also heard greenshanks in summer in North Uist. In winter the island and the Sound of Harris resound with the barking calls of barnacle geese, the honks of greylags, the whistles of drake wigeon and the lovely calls of whooper swans. Barnacle geese also occur on the Monach Islands, some five miles off the west coast, and these are essentially islands of shell-sand, which are a National Nature Reserve. The nesting birds here include skylark, wheatear, meadow and rock pipits, mallard, tufted duck and coot as well as little and Arctic terns, black guillemots, snipe, dunlin and redshank; hen harrier and lesser yellowlegs have also been reported on the islands. The mammals include grey and common seals, woodmice, shrews and rabbits.

I have been to Benbecula in winter and summer. I first landed there by air, sweeping in over a low hummocky island with a mosaic of lochans and countless inlets from the sea. A herd of whooper swans rose from a loch near the airfield, bugling in succession as they made their way towards the low horizon. In contrast the summer air is full of lark songs and the twittering notes of twites while the machair nods with golden pansies. Benbecula – a name of ill-omen to airmen in the last war fearing a posting to the back of beyond – is flat and has been called a 'pancake of an island pockmarked with water holes, lochs and lochans of every size and design'. It is also subject to mighty winds, and winter gales seem almost continuous. It was in a cave here that the Young Pretender hid in June 1746; he sheltered with a companion below the highest hill – Rueval – where I once saw a buzzard and a hooded crow perform a remarkable series of aerial evolutions high up in the air.

South Uist is a long divided island with mountains and fjords and moorland on the eastern side, where Hecla and Bheinn Mhor push up their summits almost to two thousand feet, and vast tracts of machair on the western coast. Between the two regions there is a no-man's land of lochs full of trout and salmon, and much interesting wildlife. The watery part of South Uist I found a real haven for wildlife. There are greylag geese on Loch Druidibeg in the summer and I have watched both species of diver on the island as well as little grebes, herons, mallard, teal, wigeon, shoveler, tufted duck, mergansers, mute swans and coot and there are terns – both Arctic and common – several species of gull, dunlin, redshank and snipe. Eagles can be seen soaring overhead and I once watched a pair of peregrines disporting themselves in the sky, diving and beating

234

upwards in sheer abandon. South Uist is very much the island of the machair and the lochans, and incidentally the rocket range, but I prefer to remember the living carpet of buttercups, clovers and pansies and among their flowering heads the tiny grey and fluffy balls of the eider ducklings or the yellow and black babies of the shelducks, making their way towards the sea, whose distant murmur gave a sense of infinite peace and isolation. Yet I knew that many of those tiny ducklings would fall prey to the predators that shared the island with them.

About seventeen miles off Gallan Head on the west coast of the Isle of Lewis are the seven main islands of the Flannans, or Seven Hunters, and their attendant skerries. There is an air of mystery about the Flannans. Here there is a lighthouse around whose base a great wave may once have swept three lighthouse keepers off the cliff and to their doom. The islands have a strange reputation and are difficult to land on. The Flannan Isles have grass on their tops, and Lewis crofters once used them for grazing. In 1824 Dr J. MacCulloch described them as 'like a meadow thickly enamelled with daisies'. T. S. Muir in 1885 wrote of Eilean Mor, the largest of the islands; 'on all sides it rises in mural precipices and vertical steep turfy slopes at once from the sea'. Ten of the islands and stacks have some vegetation on them. On Eilean Mor it is mainly *Holcus* with sea pink and sea milkwort. I first approached the Flannan Isles on the early morning of an autumn day – strange humps of rock with Eilean Mor like the prow of a ship sinking by the stern, and crowned with its lighthouse, set amongst a group of small stranded whales of islets. There were many fulmars off the cliffs and I saw both storm and Leach's petrels at sea. The latter breeds especially on Eilean Mor but breeding has also been proved on the other islands – Eilean Tighe, Roareim, Eilean a Gobha, Soray, and Sgeir Toman. Storm petrels also nest on the Flannan Isles as well as shags, eiders, oystercatchers, greater black-backed and herring gulls, kittiwakes, razorbills, guillemots and black guillemots, puffins, rock pipits and starlings, according to an account written by A. Anderson and others after a visit that they had made a couple of years before my own.

Even farther out to sea from the Flannan Isles is St Kilda – another group of islands and stacks, seven in all, of which Hirta is the largest; then there are Soay, Boreray, Dun, Stac an Armin, Stac Lee and Levenish. My first view of St Kilda, as of many islands, was from the air when I saw in front of me the island of Hirta – a dramatic half circle of hills – grey-green like a volcano that had blown away its nearer half and then filled with water; a shallow slope lay encompassed by the fanged island of Dun on the lefthand side and

235

the precipitous rocky nose of Oiseval on the right. As we came closer I could see the sharp triangle of The Cambir falling into the sea above a welter of stacks and partially blocking out the island of Soay beyond. Slightly to the right was the flat rise of Conachair – 1,397 feet above sea level – and down below me in the foreground the stack of Levenish, guarding the entrance to Hirta's Village Bay. The sky above was cerulean and the sea below navy-blue. Three miles beyond and in the hazy distance I could see Boreray and the two sea-girt rock pinnacles of Stac Lee and Stac an Armin. Here I was flying on the same latitude as the southern tip of Greenland but St Kilda bore an almost benign and welcoming air. Martin Martin, the much travelled island-goer, wrote in 1697 of Hirta 'the hills are often covered with ambient white mists' and the group certainly forges its own weather. It was just as Martin described the island that I saw it one day from the sea as I looked up at the great cliff of Conachair – the highest sheer sea-cliff in Britain – while myriads of fulmars circled above my head beneath the wreath of grey-white cloud that capped the island of Hirta. Rock pipits flew out from the rocky shore to circle the boat several times before flying back to the cliffs. At sea there were Arctic and great skuas, Manx shearwaters and a couple of sooty shearwaters, gannets and countless auks as the birds swarmed and flew about on the affairs of the day.

The island now seemed forbidding and somewhat awesome as we stopped off Village Bay and scanned the green-grey slope with its row of nineteenth-century cottages once lived in by the St Kildans – expert rock climbers who extracted oil from the fulmars and gathered the gannets and puffins but finally evacuated the island in 1930. I looked with interest at the primitive drystone beehive structures known as cleitts in which the villagers stored their crops, peat and the seabirds taken on their wildfowling expeditions. The whole group of islands is now owned by the National Trust for Scotland and leased as a National Nature Reserve to the Nature Conservancy Council; there is also a small garrison which plots the rockets fired from the range in the Outer Hebrides. Much of the Village has now been restored by members of the Trust. Hirta is of volcanic origin – very grassy and in summer bright with sea pinks, moss campion, primroses, roseroot, scurvy grass and even honeysuckle. Great ocean swells beat and crash up against the cliffs where kittiwakes and auks and fulmars nest in thousands, for St Kilda forms the leading seabird breeding station in Britain. Seals frequent the sea caves and sheep of the old St Kilda breed graze on the island among puffins, wheatears, meadow pipits and starlings. There are St Kilda wrens with sweeter, more far-carrying songs than those of mainland birds, and field mice but the unique house mice are extinct. Manx shearwaters and Leach's petrels nest on Hirta and storm petrels are abundant.

The island of Soay is a blunt but steep pyramid of gabbro with attendant sea stacks – Soay Stac and Stac Biorach – the latter known to Martin as the 'Mischievous Rock' – with its large guillemot colony. On the green turf live the famous Soay sheep 'untended and unimproved for more than a thousand years' and described by James Fisher as 'living fossils'. There are both dark and light types of sheep on Soay as well as breeding Leach's and storm petrels, wrens, gulls and the ubiquitous rock pipits, which live amongst the small goat-like, very agile sheep. From Soay I travelled out to Boreray and the stacks. Stark black precipices fell from Boreray into a great creamy swelling ocean that surged and ebbed around the cliffs, rock towers, pinnacles and stacks – one of the greatest natural spectacles anywhere in the British Isles. I looked up at the great east face of Boreray with its transverse cracks and ledges lined with fulmars and gannets – white splashes on the dark rock slabs which echoed with harsh raucous seabird choruses. Above lay a wreath of mist getting darker as the minutes went by, turning Boreray into 'a smoking anvil'. I saw sheep grazing on the high banks of turf, and the sea below seethed with fulmars, shearwaters, gannets and auks. Again tiny rock pipits flew out to greet the boat and flew back again. Off the western precipice of Boreray is the phenomenon of Stac Lee – a pillar of overhanging gabbro rising 544 feet from the sea – an awesome natural pillar whose young gannets or gugas were plundered by the old St Kildans in August and September, and which the late Sir Julian Huxley called 'the most majestic sea rock in existence'. As I passed through the gap between Boreray and Stac Lee the air above was full of gannets – white clouds of birds that were nesting on the two islands. To the north is Stac an Armin – a canine tooth of a stack compared to the incisor of Stac Lee. At 627 feet it is the highest sea stack in Britain and again gannets were nesting on it in hundreds; altogether nearly sixty thousand pairs of gannets nest at St Kilda. On the base of Stac an Armin about the year 1840 the last British great auk was done to death. As the boat turned away towards the Flannan Isles, Sula Sgeir and North Rona I saw wheatears, turnstones, dunlin, a sanderling and a snow bunting flying past the bow. There is a great feeling of mystery about St Kilda, with its dark forbidding cliffs, fanged pinnacles and craggy skerries, its mists and persistent drizzle, its wild seabird voices and gentle moving moan of seals, which even sun and blue sky and grassy flowery slopes do not easily dispel. The great buttressed cliff of Mullach Bi, clothed in woodrush, falls sharply from Hirta into the sea and from its top range views eastwards to Lewis and Harris, to the Uists and the ridged Cuillin of Skye. Beauty and remoteness are here set in a frame of space and timelessness of such dimensions that they are almost too majestic and awesome to bear.

It has also been my privilege to see Sula Sgeir – a half-mile long islet like a surfacing submarine, with a cave running through its middle. It is a bleak bit of land, that T. S. Muir called 'the sea-rock of the sulaire or solan goose', and Lewismen from the Port of Ness have visited it in most years to take a crop of young gugas or gannets in September; it carries a colony of about nine thousand pairs. Muir also thought that 'it presents a very naked and repulsive appearance'. It lies twenty-five miles north of Lewis and my first close view of it changed my reaction from afar when it seemed just a bare uninviting rock. I could see the gannets massed on the cliffs of the promontory of Creag Trithaiga as well as many fulmars, kittiwakes, some auks and the usual rock pipits. At night the island in summer rings with the nocturnal purring and 'chicker' notes of storm petrels and the guttural churring and 'wer-kutawuka' calls of musky Leach's petrels. Robert Atkinson, after spending nights on Sula Sgier, reported that all night he listened to 'the wild dashing and outcry of petrels'.

Twelve miles away and visible on the horizon on a clear day is North Rona, forty-four miles north-north-east of the Butt of Lewis and the most northerly of the Outer Hebrides. When I finally sailed to see Rona my cup of appreciation of island remoteness and isolation was full to overflowing. Both Rona and Sula Sgeir form a National Nature Reserve. North Rona is about three hundred acres in size, with sharp cliffs, and rises to the high ridge of Toa Rona at 355 feet; it seemed a fresh green in appearance owing to its cover of chickweed, silverweed, clover, curled dock and grasses – bents, Yorkshire fog and sheep's fescue. The flat peninsula of Fianuis to the north, somewhat richer botanically, attracts numbers of bird migrants as well as breeding Arctic terns and black guillemots, three-spined sticklebacks and grey seals. As I slowly drifted below the chapel and cell of St Ronan I counted several hundred seals hauled up on the green slopes above a grey swelling sea. The west cliff of Rona was precipitous and white with the droppings of seabirds. The island is the summer home of Leach's petrels – serenaders of the island night – and great skuas sweep over on broad brown wings. In 1933 Malcolm Stewart wrote that no one could fail to find immense pleasure in meeting North Rona for the first time and this is a sentiment that I share. I shall recall with pleasure that first sight of the island – a dark hump on the grey horizon of swelling sea – and the softly rounded bosom of Toa Rona to the right. Finally, among these remote rocky outcrops of the ocean of my acquaintance there is Foula, seen only to my regret from the air. This is the remotest of the Shetland Isles, lying sixteen miles to the west of Mainland and thirty-five miles north of Orkney. It is graced by immense cliffs. The Kame at 1220 feet is the second highest seacliff in the British Isles – a quarter of a mile of ocean precipice. There are still worked crofts but the island is often cut off by winter storms. It is the 'Bird Island' and here in summer great and Arctic skuas, Manx shearwaters and twites breed as well as larks and wheatears and many kinds of seabird. It was the setting for the film *The Edge of the World* made in 1936.

There is much that could be written about Orkney and Shetland – greatly contrasting groups of islands as they are. Orkney is green and gentle in appearance, formed from the newer, softer Old Red Sandstone, while Shetland consists of metamorphic rocks, grits, gneiss, granite and in the west Old Red Sandstone. The soil of Orkney is fertile, with sandy loams and good pasture, and it is chequered with dispersed farms and cottages, prosperous lands and fat cattle, pasture and gently rolling hills, large shallow lochs, and cliffs and stacks of Old Red Sandstone like the Old Man of Hoy. There are more than sixty islands some of which were first inhabited more than five thousand years ago. Skara Brae on the island of Mainland is an astonishing prehistoric village and there are also the chambered burial mound of Maeshow – Eric Linklater called it 'the master-work of all, magnificent in construction and unique in its truly megalithic grandeur' – the Ring of Brodgar and many cairns. This is the land of legend and the Sagas, and there are many fine bird islands – North Ronaldsay with its rare bird migrants, Papa Westray with its skuas, Eday with its cormorants and fulmars, Gairsay with its seals and seabirds, Sule Stack with four thousand pairs of gannets, and the west cliffs and inland burns, lochans and scrubby birches, rowans and alders of Hoy. There are also Eynhallow, used for ornithological research by the University of Aberdeen, especially into fulmars, and Copinsay two miles south-east of Point of Ayre, bought in 1972 as a memorial to my old friend James Fisher, managed by the RSPB and holding more than ten thousand kittiwakes, nine thousand guillemots and seven hundred fulmars. At Marwick on Mainland the RSPB have bought just over a mile of cliff, with large numbers of breeding razorbills, guillemots, kittiwakes, fulmars and smaller numbers of puffins; this site is close to where the cruiser *Hampshire* went down in 1916 with the loss of Lord Kitchener and nearly all on board. Elsewhere in addition to the cliff seabirds

Orkney can boast great and Arctic skuas, golden plover and dunlin, hen harriers and merlins, red grouse, curlew, snipe and ravens. Eiders and shelduck, teal, wigeon, pintail, shoveler, scaup and tufted duck breed on some of the moors and lochs, and there are also twites and short-eared owls. My visits to Orkney were all confined to Mainland where I have watched those handsome birds – the red-throated divers – and many terns. The RSPB has several moorland reserves in Orkney. The chief town is Kirkwall – a modern town with its St Magnus Cathedral, Bishop's Palace, tall trees from which I heard the moan of collared doves and the songs of robin, wren, hedgesparrow and starling, and a pleasant harbour.

The Shetland Isles consist of over a hundred islands of which about seventeen are inhabited. They lie some 220 miles west of Bergen in Norway and have little in common with Orkney except their early history. The largest island is Mainland, some fifty miles long and twenty miles across at the widest part. Close by are the islands of Bressay and Noss to the east and Whalsay, Yell, Fetlar and Unst to the north-east. Nesses, stacks and voes are the terms that are liberally scattered over a map of Shetland, representing headlands carved from the sea, isolated rocks at sea, and long inlets cut so deeply into the land that no Shetlander is ever more than three miles from the sea. Sea and ice have left their marks on the harder rocks of Shetland – the gneiss, schists and granite. There are hills of heather and grass, vast tracts of peat and many lochans. Good soils are scarce and Shetland looks more like the Scottish High-lands or Norway than Orkney.

I have reached Shetland by air to Sumburgh Airport and sailed to Lerwick on the *St Clair* from Aberdeen. In winter I have seen Sumburgh swept by great gales charged with salt and by cold wet winds blowing from an infinity of grey skies. One February the aircraft from Dyce gave me a fine bird's eye view of Fair Isle after leaving Kirkwall, and as I looked down I could see the tiny grey-white crosses of fulmars soaring and weaving intricate patterns above the island's cliffs and stacks. Soon the aircraft was dropping below the headland ridge of Sumburgh onto Mainland. Five minutes after I had landed a blizzard struck savagely from the north and the black and greyish hills quickly disappeared under a carpet of snow. All the way to Lerwick sheep with snow-packed fleeces huddled for shelter in the lee of peat cuttings and burns or behind stone walls. Lerwick itself suddenly appeared, ghostly and grey on its hillside, with low white clouds scudding across a sky of astonishing black-ness. The whole world was drained of colour.

In the Sound of Bressay beyond Lerwick's large harbour eiders, gannets, black guillemots, or 'Tysties', in winter plumage, and shags were swimming or feeding. Gulls circled above the quays – chiefly herring gulls but I also saw common, black-headed and immature glaucous and Iceland gulls. Across the black water came the deep throaty barks of great black-backed gulls riding high on the choppy Sound, and outside my hotel room two herring gulls called antiphonally. On the outskirts of the town I found a loch and on it the remaining walls of Clickhimin fort or 'broch', where carved stones depicting footprints were found, and built perhaps as a permanent fortress by small clans of Celts seeking a defence against the marauding Picts from the south. On the water around the broch I could see tufted ducks and goldeneye and around the edge scores of ravens and hooded crows, which started to take off noisily over the town on a fresh foraging expedition. On the steep slopes above Commercial Street and the lodberries – the pile-supported houses that jut out into the sea – was a walled garden with a few salt-blasted sycamores, from which came bursts of exotic song; here just over four hundred miles from the Arctic Circle the Asiatic tropical collared doves seemed at home in the snow. Shetland starlings sang vigorously from wires and posts, imitating curlews or calling 'skreer' in their special island way.

One dull winter afternoon I went across the island to Scalloway and here beneath the seventeenth-century castle of Earl Patrick Stewart and on this western voe parties of long-tailed ducks were displaying on the sea. The drakes dashed about and around the females with their tails cocked up and a great deal of bowing and musical yodelling and 'cahooing', which travelled over the water with a touch of wild mystery and charm. I also came across a small herd of whooper swans on Trondavoe Loch; in Shetland the musical bugling notes of these swans have been likened to the clearing of anchor chains and the singing of capstan shanties. As the days went by the snow melted, a slight warmth came into the sun's rays and I explored the Norse steatite quarry with some of the nearly completed soapstone pots still adhering to their mother rock. Near Fladdabister – a jumble of white-walled cottages, thatched black houses, roofless shells and crofts – I disturbed a large flock of snow buntings and at least fifty twites on a hillside plot. At the southernmost tip of Mainland is the Neolithic, Bronze and Iron Age, Norse and medieval settlement of Jarlshof – an assortment of pits, earth-houses, brochs and farmhouses grouped around 'de Laird's House' – a medieval manorhouse – or Jarlshof as it has been known since Sir Walter Scott described it in *The Pirate*. Once under piles of windblown sand it was excavated and lay revealed as one of the greatest archaeological treasures in the British Isles, and in 1965 I directed a film about it for the BBC. As I walked between the ruins and across the neatly mown grass swards that winter day, my last impressions of Shetland were of fifty rock doves wheeling above my head, of three fulmars cackling from an ancient wall and of a redshank and a dozen trilling turnstones that rose up from the grass in alarm.

I have also seen Shetland in summer and listened to the evening chorus – albeit a thin one – of blackbirds above the Sound of Bressay. I have walked across that island to see Noss with its 4,300 pairs of gannets breeding on the Noup, many other seabirds and fierce skuas. I have been regaled by the long calls of common gulls on the lochsides and grassy slopes near the sea and even on the heather moors. I have listened to the scream of terns around the Broch of Mousa. Unst and Fetlar with their bogs of cotton grass and lochs held red-throated divers, sandpipers, mergansers and whimbrels whose bubbling succession of fast notes, rising and falling in pitch, is for me one of the loveliest of all wader 'songs' – standing comparison with the display of the stone curlew in Breckland. There were red grouse and golden plover on some of the hills in Shetland and there was a sycamore wood where rooks and woodpigeons nested and even long-eared owls, blackcaps and song thrushes. There were still some red-necked phalaropes too. Curlew, lapwings, oystercatchers and ringed plover were frequent on the islands. In summer Shetland is a paradise for birds; it is a land of dark heathery moors, green islets and the 'simmer dim' – that long Shetland summer night that compensates for the wild winters. Daylight here is a more northern light quite unlike the opaque halations of Orkney or the scene in the Outer Hebrides. There are strange reefs of gneiss rising from a black sea – Hoo Stack, Muckla Billan, Muckla Fladdicap terminating in the Out Skerries which push towards the island of Fetlar so long associated with the snowy owl. There are the cliffs of Hermaness with their great bird cities and at their foot the hauling out places of the grey seal. When I first knew Shetland it was calm and assured and these characteristics were to help the islands when they had to come to terms with Sullom Voe and the oil industry. The visitor or the tourist may have a romantic, perhaps ultra-sentimental view of Shetland but the islanders themselves have needs, and so environmental planning has been arranged to cover all the interests of local people, industry, settlements, tourism and the natural scene. The Shetlanders have never been insular in their outlook and this may well stand them in good stead in the future.

My island-going has also taken me to one of the most remote of all the inhabited islands – the Fair Isle. It is some three miles long and two wide and it lies in the distant ocean some twenty-five miles south of Fitful Head in Shetland and halfway between Sumburgh and North Ronaldsay in Orkney. The island is famous for its Bird Observatory, its exquisite knitwear, its magnificent cliff scenery which is unsurpassed in Britain and its kind and warm-hearted people. One summer I went to Grutness Pier in the south of Mainland in Shetland and barely a mile from Jarlshof to board the Fair Isle mail boat *Good Shepherd*. After dropping her incoming cargo of lobsters we sailed on the three-hour voyage to the island. The boat tossed, swung and fought her way through the dreaded Sumburgh 'roost', that turbulent area of water where Atlantic rollers meet the swells of the North Sea. We finally entered the North Haven of Fair Isle, passing great cliffs and stacks on either side; in front we could see the Observatory, the pier and the extraordinary sloping sandstone mass of the Sheep Craig, joined to the slopes of Vaasetter by a knife-edge of crumbly rock and accessible for men and sheep only from the sea. I stayed at the Bird Observatory and here a Fair Isle wren – of a special race, by the way – bombarded me with its powerful and distinctive song from the roof. Amongst the buildings were numbers of starlings whose young had just left their nests in the drystone walls. The commonest birds here were the twites, which sang and called to each other all day. I saw quite a few migrant warblers and waders, several collared doves and a woodchat shrike. Indeed Fair Isle is famous for its rare birds and has more 'first records' than any other place in Britain. The stark isolation of the Isle makes it a cross-roads of the air for millions of bird migrants and the list of rarities is truly impressive; in 1977 alone the visitors included Pechora pipit, lanceolated warbler, booted warbler, scarlet rosefinch and yellow-breasted bunting.

The island belongs to the National Trust for Scotland, which acquired it from the ornithologist George Waterston in 1954. The male islanders were busy lobster fishing, working their crofts, weaving textiles; some of them are noteworthy as expert field naturalists in their own right. The north part of the island is high sheep grazing and I have stood on the top of Ward Hill high above a bank of white cloud that overshadowed the island and deprived the islanders of the sun that I was enjoying. The crofts are scattered over the lower land in the south and from the air the scene is one of a score of tiny white or grey houses lying at right angles to the white strips of the two parallel trackways running north and south. Crops and vegetables are raised in small four-sided stone enclosures known as 'planti-crubs'. The growing season stretches from 'da voar' – the spring sowing – to 'da hairst' – the harvest of ryegrass, oats or potatoes. Sheep and some cattle graze between Meoness and the west coast. The cliffs, on which spectacular and brave rescues have often been made by the islanders, are truly mighty, rising to six hundred feet and contemplating a host of contorted pinnacles, shafts, arrètes and reefs of rock of every size and shape. On the north coast the Old Red Sandstone provides the dramatic nameless stack, Cathedral rock and the skerries – land of kittiwakes, ravens and grey seals. Teeming colonies of kittiwakes, fulmars, shags and auks are balanced on the ledges, and black guillemots whistle hollowly by the great blow-hole of the Kirn o'Skroo. There are rabbits on the island – grey, black, albinos, white-collared and some with long silky coats like Yorkshire terriers; Charles Lagus filmed some for my portrait of Fair Isle, *Island Home*. It is probable that some of this stock was descended from pets kept once by the lighthouse keepers. I also saw bright fox-coloured wood mice – large and distinctive and probably twice as heavy as those I had seen on the mainland of Britain.

For some of the days a fine sea-mist enshrouded the island – not thick enough to drain the sun of its warmth but enough to threaten ships at sea, and from the southern tip of the island the fog horn kept up its constant warning. From the school playground came the laughs and shouts of children – seven in 1963 when I made the film but by 1977 there were over twenty children of school age. The population is eighty at the time of writing; of these two men are weaving and only one woman knits the Fair Isle all-over sweater, one of which is my very proud possession. There are a Post Office, the school, of course, a village hall, a church and a chapel, and an airstrip from which it is a quarter of an hour's flight by Islander aircraft to Sumburgh. The people are deeply religious, and there is no public house. On misty days I went out on a lobster boat – one of the Fair Isle yoals – and explored every crack and geo in the sea cliffs. Once we penetrated a deep sea cave whose cliffs rose sheer hundreds of feet above me, and then at the end of the tunnel was an almost vertical shaft rising towards the distant light. Shags flew low past me like giant bats as I stood in the prow of the boat and kittiwakes poured down a storm of droppings from the inside of the shaft. The water in the cave was deep and incredibly clear even in the poor light. We heard a faint bleat from a ledge beneath the shaft and found a lamb, starving and matted with bird droppings, which had fallen down the shaft and lived; its mother was killed in the fall and lay nearby. I gathered up the lamb, acrid with ammonia, warmed it in my jacket and brought it back to land.

Far above the cliffs are heathery slopes which hold the territories of the piratical buccaneering skuas. The bonxies – the great skuas –

244

came wheeling above me, some to stoop savagely at my head or just cruise menacingly by, calling quietly 'a-er' all the time. Others sitting on low mounds a hundred yards away would start a low-level flight, sweeping in a foot or so above the ground, and press home an attack with panache before lifting up in a rush of wings six inches from my forehead – an unnerving experience. There were also the slightly built, swift, narrow-winged Arctic skuas or 'skuti-alans' like strange dark falcons, which frequently passed overhead in their fierce chase of each other, calling with mewing reedy wails. Both skuas are strangely beautiful birds but much of their time is spent pursuing gulls, terns and auks and forcing them to yield up their last meal. On the Fair Isle I went through a whole spectrum of enjoyable experiences – sharing Vaasetter with the arrogant great skuas, spending hours watching the rabbits or the puffins at their burrows, admiring the heroic seascapes, appreciating the flowering squills, sea pinks and sticky mouse ear in delicate blossom and even the duller star sedge and least willow, and being privileged to share the lives of these kind and hard-working people whose story I tried to tell in my film *Island Home*. Some of the happiest and most rewarding days of my life were passed on the island and the final moment of departure was sad in the extreme. As the *Good Shepherd* carried me out to sea from the North Haven a rock pipit launched into a last song of farewell.

Index